KINEMATICS OF INDUSTRIAL ROBOTS

WILMER EDUARDO SANZ FERNÁNDEZ

Cinemática de Robots Industriales.
1st edition printed in Spanish, October 2009.
The 2nd edition printed in Spanish, in March 2013.
Registered work in the Autonomous Service
Intellectual Property (SAPI), under No. 7542

Done the legal deposit
Legal Deposit: 252201660054 LFI
ISBN 978-980-12-8534-2
Valencia, Venezuela.

ISBN: 978-980-12-8534-2

9 789801 285342

ACKNOWLEDGMENTS

Thanks:
To God, my Lord.
To my mother (†), Josefina Fernández.
To my father (†), Eduardo Sanz.
To my wife, Norela and my son, Ángel de Dios.

TABLE OF CONTENTS

THE AUTHOR

Wilmer Sanz is Electrical Engineer, Specialist in teaching for Higher Education and Magister Scientiarum in Instrumentation. He works as Full Professor, head of the Laboratory of Robotics and Industrial Vision and head of the professorships of Robotics and Computer Architecture in Department of Systems and Automation of School of Electrical Engineering at the University of Carabobo (Valencia, Venezuela). Author of the book "Integrales, Métodos y Aplicaciones" (Integrals, Methods and Applications), he has taught courses, performed and directed numerous researches in the undergraduate level, Specialization and Master in several Venezuelan universities; and has been a consistent promoter of Industrial Robotics in the Latin American, generating free products for modeling and simulation of tasks with robot manipulators. With over 20 years of top-level teaching career, Professor Sanz offers his pedagogical vision as a contribution to learning of Robotics.

PREFACE

Often the Sci-Fi's world has been the crystal ball used by visionaries and prophets of technology. In their books, authors like Jules Verne, H. G. Wells or Isaac Asimov, have predicted with success and decades of anticipation, numerous inventions and the consequences of their apparition in the middle of human societies. However, what is the explanation of this phenomenon? Before to answer to this question, are necessaries several reflections.

"Under the Sun, nothing is new", says a book of Bible called Ecclesiastes. Instead of inventing, man rediscovers and adapts; can tread new paths and even make them, can explore new paths and even make them, but the land on which transits is old and full of stories lived by many others before him. However, it is fair to recognize the role of the pioneers, who are ahead to help their congeners, making ways and bridges that will cross entire peoples behind them.

It is worth seeing how the men seek to satisfy their need for progress and development inspired by models. Today millions of people are Christian or Buddhist, by follow the principles of Love and brotherhood shown by Jesus and Siddhartha; many others believe in the movement of Not Violence or “Ahimsa”, because Gandhi showed how use it to reach goals like Liberty; and crowds move when a leader makes them believe that a target is possible or simply useful.

Now, my personal answer to the previous question: I see to modern prophets and visionaries as road designers, that the pioneers choose walking. I believe the clairvoyance is based on knowledge of the needs, in the desire and efforts to discover possible solutions and satisfactions. In this form an idea about “what is possible” becomes "what will be” or “what is” inside of a futurist context; and according such idea, sooner or later, a pioneer takes it like inspiration and discovers the way to make it becomes into reality, smoothing a road, discovering a path.

Inspired by these thoughts and realize that robotics is a broad field to work in a research promisingly fertile, I plan to facilitate the formation of enthusiastic and students about the modeling and analysis of robots, al level of professionals in Electrical Engineering, Electronic Engineering and Mechatronics. Because these reasons the kind of studied robot in this book is the called Industrial Robot (IR), used in the industries for manipulation of tools and pieces with the goal to facilitate and optimize the human work; not the explorer, humanoid or insectoid.

The book is organized in this form: the first chapters include the exposition of a hard theory in appearance but really is easy for everyone with knowledges of matrix algebra and trigonometry. Next, all other chapters consist in the use of that theory in the kinematic analysis of more used configurations of Industrial Robots, others less used configurations and historical robots.

My hope is that this book will be useful to illustrate the most popular procedures used in the kinematic modeling of IR, in a clear and pedagogical form in comparison to other books of robotics.

Finally, I use these lines to thank the Light that makes me live for a higher purpose and also to the fellows who encourage me to do it in ways like this: my students.

"May Eru bless my work and amend it!"
El Silmarillion. J.R.R. Tolkien

CHAPTER 1

INDUSTRIAL ROBOTS CONFIGURATIONS

"We can do no great things, only small things with great love"
Agnes Gonxha Bojaxhiu.

The combinations of the revolute (R) and prismatic (P) joints, in the first three axes, result in eight possible configurations for the Industrial Manipulators. Designers and manufacturers have preferred four of them, because of versatility and controllability of the produced types. In reality, one of these combinations has generated two kinds of different robots and the most specialists identify five configurations how more accepted in the industries. Table 1.1 shows all configurations including the five previously announced.

Table 1.1 Configurations and types of IR

Configuration	***Industrial Robot***
PPP	Cartesian
PPR	
PRP	
PRR	
RPP	Cylindrical
RPR	
RRP	Spherical and SCARA
RRR	Anthropomorphic

The figure 1.1 shows representations of the most common configurations.

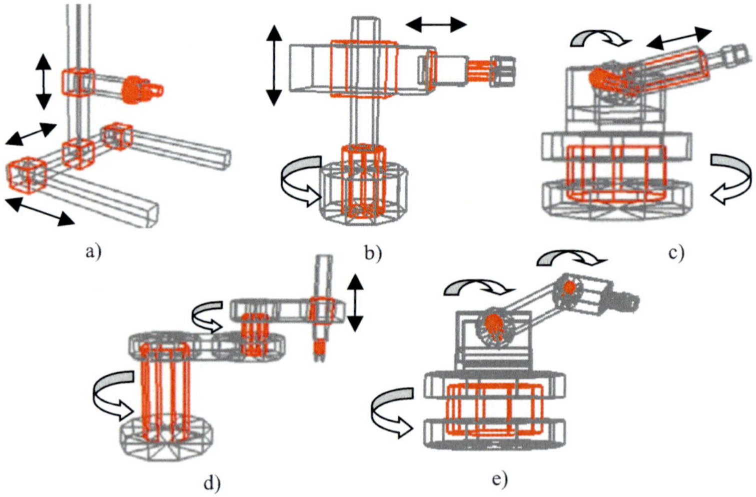

Figure 1.1 More popular configurations of IR. a) Cartesian. b) Cylindrical. c) Spherical. d) Toroidal or SCARA. e) Angular or Anthropomorphic

1.1 CARTESIAN ROBOT

With three prismatic joints, perpendicular to each other, this may be the 2nd or 3rd kind of IR most frequently used. It is notable for to have easy controllability, precession and to work with speeds above the average of other types of IR (it is because they have no revolute joints, which are sensitive to pseudo-forces in circular movements).

Useful in palletizing and transport tasks of parts in storage racks, the robot executes the linear movements corresponding to a coordinate system XYZ.

For three linear movements with L_1, L_2 and L_3 magnitudes, its Work Volume or Workspace (all reachable points by the final link of robot) has a parallelepiped form (figure 1.2) and may be calculated with the equation 1.1.

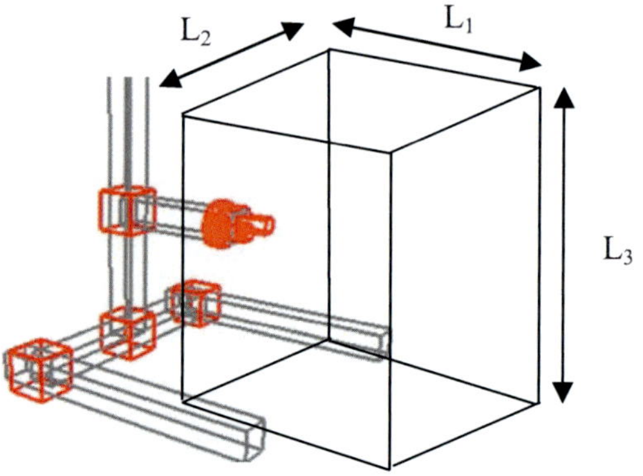

Figure 1.2 Work Volume of Cartesian Robot

$$V_{cartesian} = L_1 L_2 L_3 \tag{1.1}$$

1.2. CYLINDRICAL ROBOT

This configuration executes movements corresponding to a cylindrical coordinates system, where there is an angular variable and another two linear, mutually perpendicular, to reach any point of the space. Like you can see in the figure 1.3, only the points inside the space between two cylinders, belongs to the Work Volume.

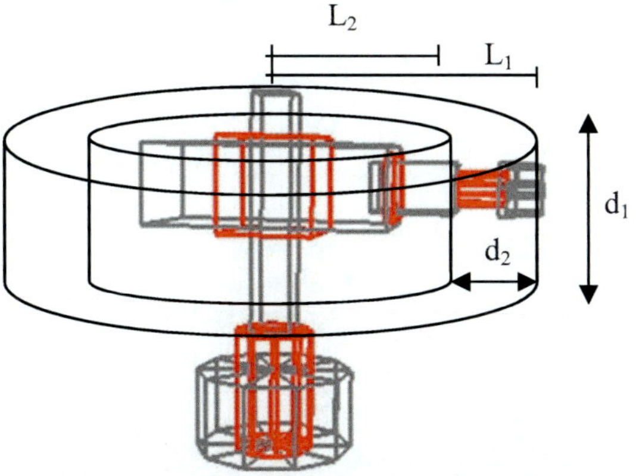

Figure 1.3 Work Volume of Cylindrical Robot

According to the figure 1.3, if $d_2=L_1-L_2$ (L_1 and L_2 are the maximum and minimum turning radius of the robot) then the Work Volume may calculated by means of the equation 1.2.

$$V_{cylindrical} = \pi(L_1^{\,2} - L_2^{\,2})d_1 \tag{1.2}$$

1.3. SPHERICAL OR POLAR ROBOT

Like the Cylindrical Arm, this configuration has been designed to reproduce the movement of a coordinate system, the spherical in this case; with two angular movements (equivalent to the latitude and longitude used in navigation) and another linear (figure 1.4).

The equation 1.3 allows determining the Workspace from the observed parameters in the figure 1.4 (for turning radius R_1 and R_2, $d_3=R_1-R_2$).

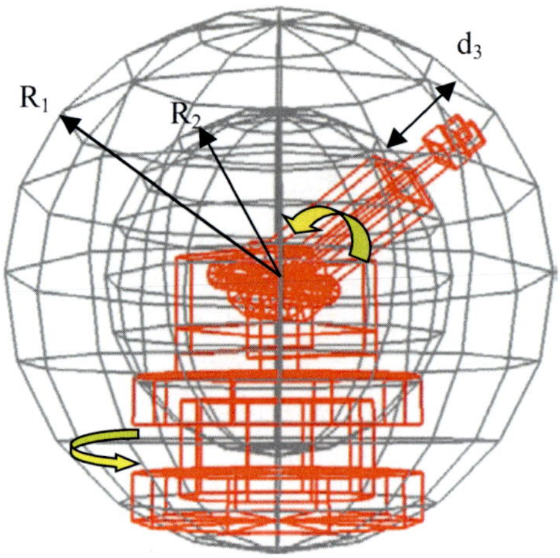

Figure 1.4 Workspace of Spherical Robot.

$$V_{spherical} = \frac{4}{3}\pi(R_1^3 - R_2^3) \qquad (1.3)$$

1.4. TOROIDAL OR SCARA ROBOT

This is the second of RRP named models in the Table 1.1. Unlike Spherical Robot, where the joints 1 and 2 are orthogonal, here they are parallels.

SCARA is the acronym for *Selective Compliant Assembly Robot Arm* or *Selective Compliant Articulated Robot Arm*; apparently both denominations are equally valid or accepted. The word "Selective" refers to the maneuverability of the robot in the XY plane and its rigidity in the third joint. It makes it unsuitable for working in planes involving the Z direction.

However, the original identification of robot was *Selective Compliance Assembly Robot Arm*, such as decided its creator Hiroshi Makino, professor of Yamanashi University in Japan, at beginnings of 80's.

The SCARA robot was conceived as an alternative to work in parallel plans, cheaper, more single and faster than other robots. Indeed, it has proven to be a very appropriate choice in packaging tasks, electronic assembly and in general for any task that requires handling parts on a table or work surface.

The figure 1.5 shows the Work Volume of this robot, which is similar to a ring or toroid (like a donut). The depth or third dimension of this volume (d_3) is implemented by the prismatic joint when it is moved crossing the planes formed for circles that remember the geocentric model with epicycles of Claudius Ptolemy (figure 1.6).

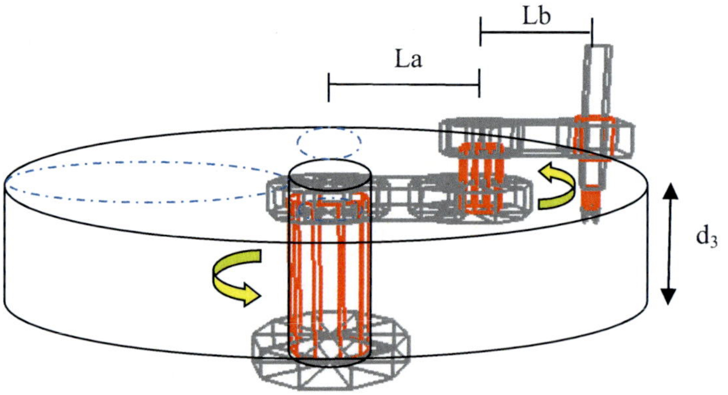

Figure 1.5 Workspace of SCARA Robot

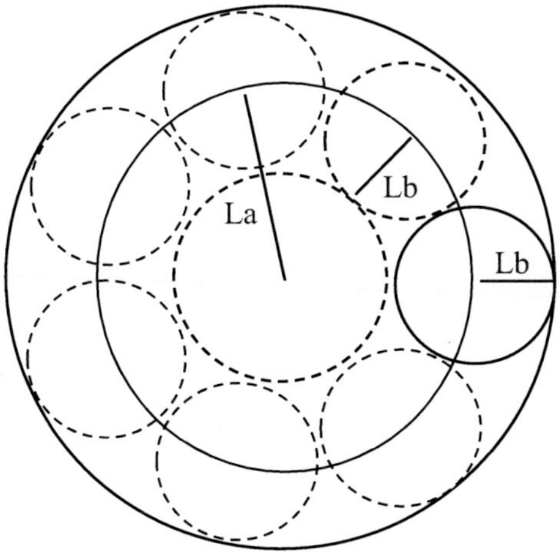

Figure 1.6 Work area of SCARA Robot

To determine the Workspace:

$$V_{SCARA} = [\pi(La + Lb)^2 - \pi(La - Lb)^2]d_3$$

$$\boxed{V_{SCARA} = 4\pi LaLbd_3} \quad (1.4)$$

1.5. ANTHROPOMORPHIC ROBOT

The configuration RRR produces the most versatile IR. With three revolute joints, the first orthogonal to the followings, this robot is called Anthropomorphic because mimics the movements of the hip, shoulder and elbow of the human beings (figure 1.7).

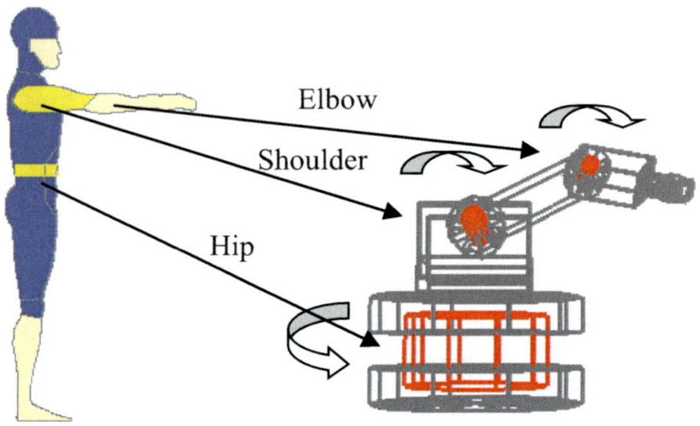

Figure 1.7 Joints of Anthropomorphic Robot

In principle, the end effector of the robot can reach all the internal points on a sphere of radius equal to the sum of the length of the "arm" and "forearm" (figure 1.8) whose volume is determined by the equation 1.5.

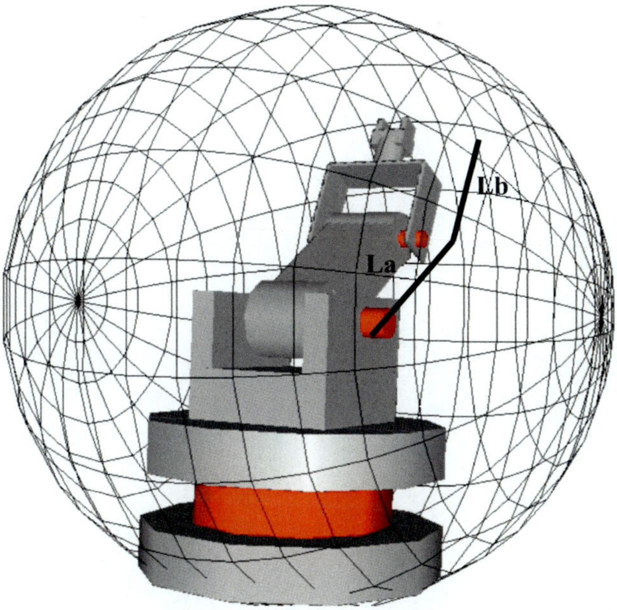

Figure 1.8. Workspace of Anthropomorphic Robot

$$V_{Anthropomorphic} = \frac{4}{3}\pi(La + Lb)^3 \qquad (1.5)$$

The Robot is useful in many applications, especially those requiring careful around obstacles and access in narrow cavities.

CHAPTER 2

REPRESENTATION OF THE SOLID'S ORIENTATION

"We have modified our environment so radically that we must now modify ourselves to exist in this new environment".

Norbert Wiener

Three coordinates are enough to determine a position into space, but a solid body is not a point, else a multitude of points with a singular geometry, which is necessary defining another concept to describe its spatial condition: the Orientation. Given a reference system O'UVW, called mobile, which is solidly attached to the manipulated object, the Orientation may be understood as its relationship with another reference system OXYZ, called fixed (see figure 2.1).

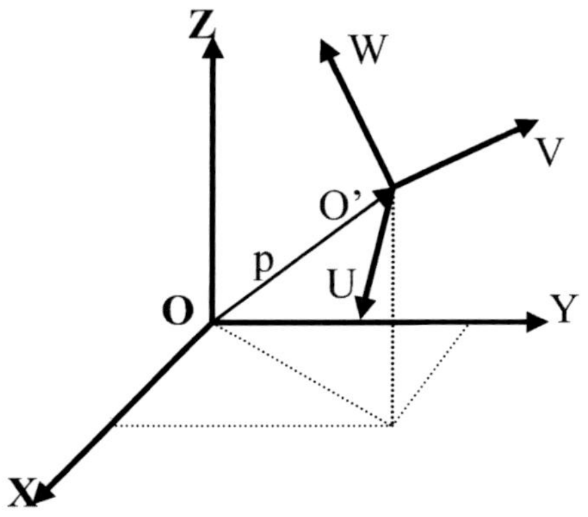

Figure 2.1 Orientation: Relationship between two reference systems, mobile and fixed.

The three axis of the mobile reference system have the following unit vectors (figure 2.2):

- *Approach* ($\vec{a}$): It is defined on the W axis and indicates how the tool faces or "attacks" to the work plane.
- *Slide* ($\vec{s}$): It belongs to V axis and determines the direction in which works the tool. The figure 2.2 shows the slide vector for an end effector with two-finger grippers.
- *Normal* ($\vec{n}$): This completes a right-handed system of unit vectors in the reference mobile frame. It is defined in the direction of U axis.

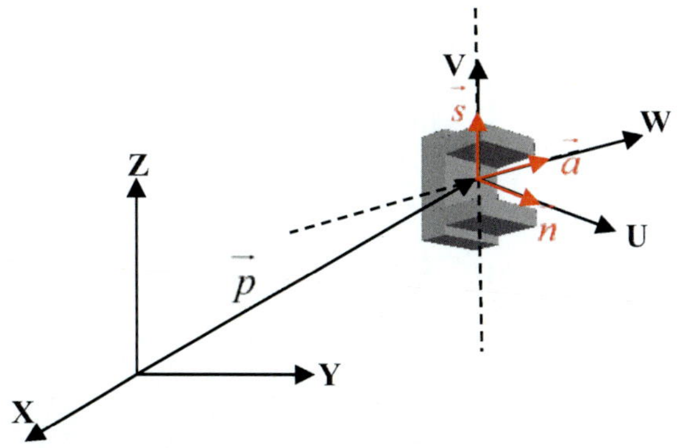

Figure 2.2 Normal, Slide and Approach vectors

2.1. ROTATION MATRICES

To introduce the problem of modelling Orientation, let us begin with an initial condition where the reference frames, fixed and mobile, have total coincidence of their axes and then we proceed to rotate the mobile system around the Z axis, in θ angle (figure 2.3).

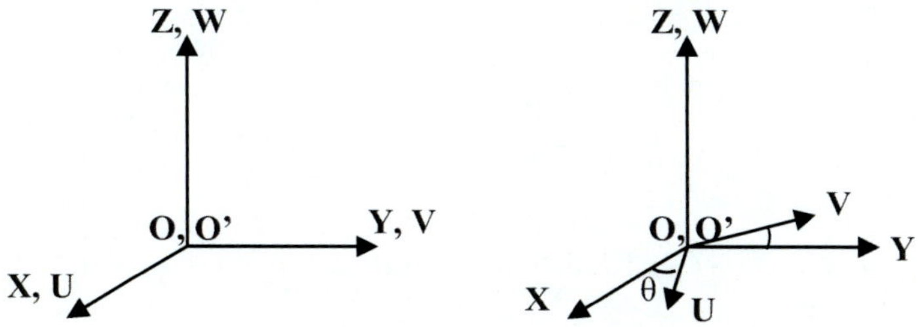

Figure 2.3 a) Initial condition of coincidence between reference frames, mobile and fixed.
b) Rotation of mobile reference frame around the Z axis.

An overhead view of the condition shown in the figure 2.3, is appropriate for analyze the problem from the XYUV plane, as is visible in the figure 2.4.

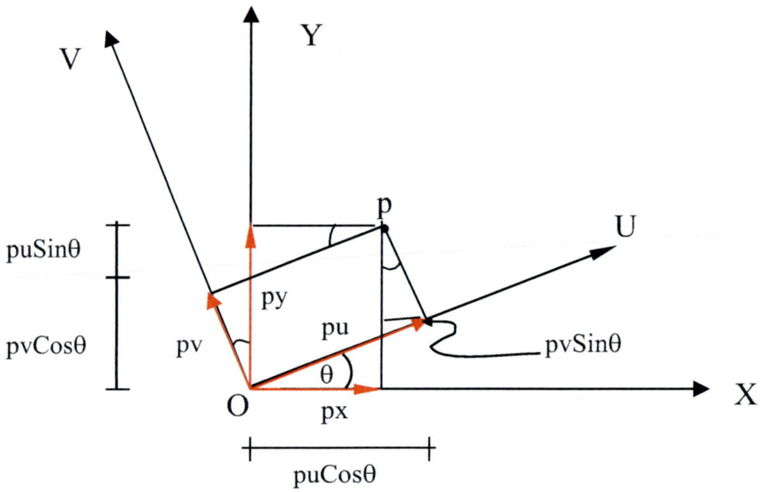

Figure 2.4 Analysis of a pure rotation around the Z axis, seen in the XYUV plane

From figure 2.4 is easy to obtain the following pair of equations:

$$px = puCos\theta - pvSin\theta$$
$$py = puSin\theta + pvCos\theta$$

You may rewrite the above equations in matrix form:

$$\begin{bmatrix} px \\ py \end{bmatrix} = \begin{bmatrix} Cos\theta & -Sin\theta \\ Sin\theta & Cos\theta \end{bmatrix} \begin{bmatrix} pu \\ pv \end{bmatrix}$$

The central matrix is known as Rotation Matrix (R). It describes the variations in the orientation mobile frame 0UV, respect to fixed frame, 0XY.

A more general expression of this matrix arises to relate their entries with dot products of unit vectors corresponding to the axes of involved systems. The figure 2.5 shows these vectors and the angles between them.

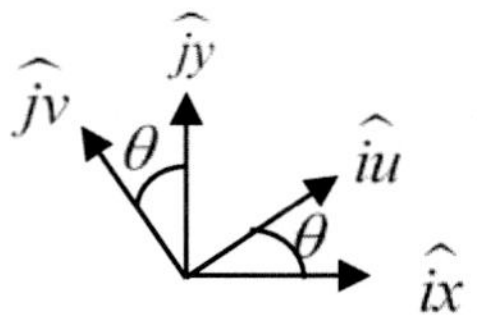

Figure 2.5 Unit vectors of XYUV plane

The dot products are:

$$\widehat{ix} \bullet \widehat{iu} = \left|\widehat{ix}\right|\left|\widehat{iu}\right| Cos\theta = Cos\theta$$

$$\widehat{ix} \bullet \widehat{jv} = \left|\widehat{ix}\right|\left|\widehat{jv}\right| Cos(90+\theta) = -Sin\theta$$

$$\widehat{jy} \bullet \widehat{iu} = \left|\widehat{jy}\right|\left|\widehat{iu}\right| Cos\theta = Cos(90-\theta) = Sin\theta$$

$$\widehat{jy} \bullet \widehat{jv} = \left|\widehat{jy}\right|\left|\widehat{jv}\right| Cos\theta = Cos\theta$$

Replacing entries of Rotation Matrix from above results:

$$R = \begin{bmatrix} \widehat{ix} \bullet \widehat{iu} & \widehat{ix} \bullet \widehat{jv} \\ \widehat{jy} \bullet \widehat{iu} & \widehat{jy} \bullet \widehat{jv} \end{bmatrix} \tag{2.1}$$

It is possible to identify a tendency of dot products inside this expression. Indeed, the dot products of first row have $\widehat{ix}$ as first factor, while $\widehat{jy}$ does the same in the second row. On the other hand the first column has $\widehat{iu}$ as second factor, and $\widehat{jv}$ doing the same in the second column. Extrapolating, we can infer new dot products with $\widehat{kz}$ and $\widehat{kw}$ (respectively unit vectors of Z and W axis), in the case of Rotation Matrix corresponding to Euclidean space (equation 2.2).

$$R = \begin{bmatrix} \widehat{ix} \bullet \widehat{iu} & \widehat{ix} \bullet \widehat{jv} & \widehat{ix} \bullet \widehat{kw} \\ \widehat{jy} \bullet \widehat{iu} & \widehat{jy} \bullet \widehat{jv} & \widehat{jy} \bullet \widehat{kw} \\ \widehat{kz} \bullet \widehat{iu} & \widehat{kz} \bullet \widehat{jv} & \widehat{kz} \bullet \widehat{kw} \end{bmatrix} \tag{2.2}$$

This equation is known as **Direction Cosine Matrix** or **Rotation Matrix** of three dimensional space. It describes the variations in the orientation mobile frame 0UVW, respect to fixed frame, 0XYZ.

2.2. CONSECUTIVE ROTATIONS

The next problem to be solved is modelling a rotation around to an oblique axis producing, any orientation in the space. As approaching to solution, first let us to study the cases of pure rotations around each axe of the fixed frame, 0XYZ (figure 2.6).

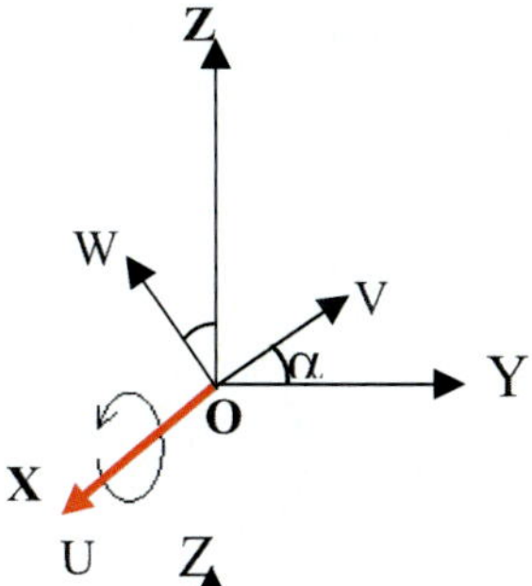

$$R(x,\alpha)=\begin{bmatrix}1 & 0 & 0\\ 0 & Cos\alpha & -Sin\alpha\\ 0 & Sin\alpha & Cos\alpha\end{bmatrix}$$

a) Turn around X axis

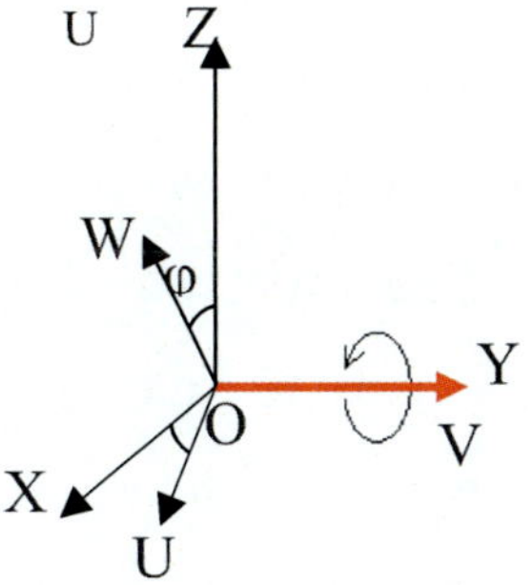

$$R(y,\varphi)=\begin{bmatrix}Cos\varphi & 0 & Sin\varphi\\ 0 & 1 & 0\\ -Sin\varphi & 0 & Cos\varphi\end{bmatrix}$$

b) Turn around Y axis

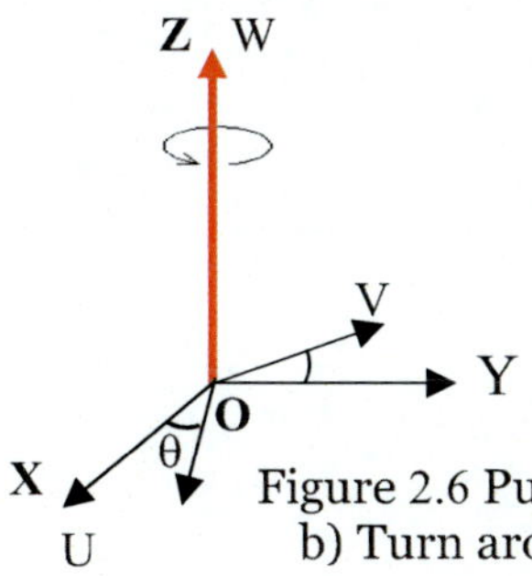

$$R(z,\theta)=\begin{bmatrix}Cos\theta & -Sin\theta & 0\\ Sin\theta & Cos\theta & 0\\ 0 & 0 & 1\end{bmatrix}$$

c) Turn around Z axis

Figure 2.6 Pure rotations: a) Turn around X axis, b) Turn around Y axis, c) Turn around Z axis

Now, according to Superposition rule, we can reproduce any orientation reached by mean of a rotation around an oblique axis (figure 2.7), through combinations of successive pure rotations, around the X, Y and Z axes.

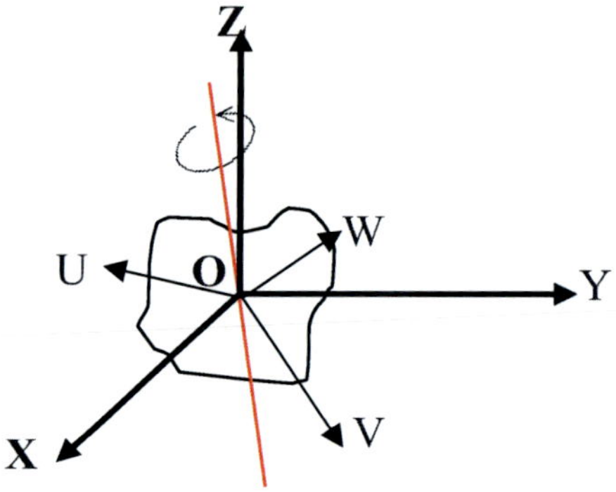

Figure 2.7 Turn around an oblique axis

The execution of successive rotations must be modelled by multiplying of respective Rotation Matrices. The order of multiplying requires being inverted (pre-multiplications) respect to the announced sequence. E.g., the rotations sequence: a ψ angle, around X, followed by a θ angle, around Y and finishing with a φ angle, around Z axis; must be written as follows:

$$R(z,\varphi)R(y,\theta)R(x,\psi) =$$

$$\begin{bmatrix} Cos\varphi & -Sin\varphi & 0 \\ Sin\varphi & Cos\varphi & 0 \\ 0 & 0 & 1 \end{bmatrix} \begin{bmatrix} Cos\theta & 0 & Sin\theta \\ 0 & 1 & 0 \\ -Sin\theta & 0 & Cos\theta \end{bmatrix} \begin{bmatrix} 1 & 0 & 0 \\ 0 & Cos\psi & -Sin\psi \\ 0 & Sin\psi & Cos\psi \end{bmatrix}$$

2.3. RPY ANGLES

The result of the above operation is well known in aviation. The angles used are known as RPY, the initials of Roll (ψ), Pitch (θ) and Yaw (φ), which corresponds to turns around XYZ axes defined to an airplane as shown in the figure 2.8.

$$RPY(\varphi,\theta,\psi)=$$

$$\begin{bmatrix} Cos\varphi Cos\theta & -Sin\varphi Cos\psi + Cos\varphi Sin\theta Sin\psi & Sin\varphi Sin\psi + Cos\varphi Sin\theta Cos\psi \\ Sin\varphi Cos\theta & Cos\varphi Cos\psi + Sin\varphi Sin\theta Sin\psi & -Cos\varphi Sin\psi + Sin\varphi Sin\theta Cos\psi \\ -Sin\theta & Cos\theta Sin\psi & Cos\theta Cos\psi \end{bmatrix}$$

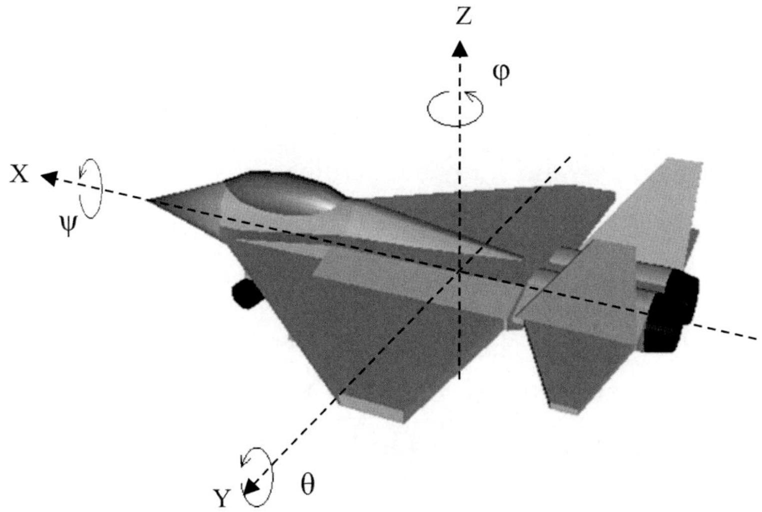

Figure 2.8 RPY angles. Image based on airplane model included in the software RoboWorks Demo 2.0

The roll consists of upward movement of a wing and lowering the other due to rotation about the longitudinal axis of the aircraft. The pitch is the declination or lifting the tip of the plane, equivalent to rotation about a transverse or lateral axis extending along the wings. Finally, yaw represents the rotation about a vertical axis.

2.4. ABSOLUTE, RELATIVE ROTATIONS AND EULER ANGLES.

When consecutive rotations are realized as the case of angles RPY above described, the corresponding transformation is obtained by means of multiplying rotation matrices. Since **the product of matrices is not commutative**, is necessary preserve the order of multiplications. The figure 2.9 clearly shows how two completely different conditions are obtained when the sequence of rotations is inverted; and how we must doing pre-multiplications in both cases in order to obtain the equivalent transformation. **The pre-multiplications are written from right to left, i.e. the last executed must be written at beginning, while the first executed must be written at end**.

In the figure 2.9, in both cases of the example, the rotations are made around the XYZ axes of fixed frame. These are called **Absolute Transformations**. Unlike, in the figure 2.10 the rotations are realized around the UVW of mobile frame, calling them as **Relative Transformations**. We can't assume a priori the above procedure to obtain the equivalent transformation. How must be done in this case?

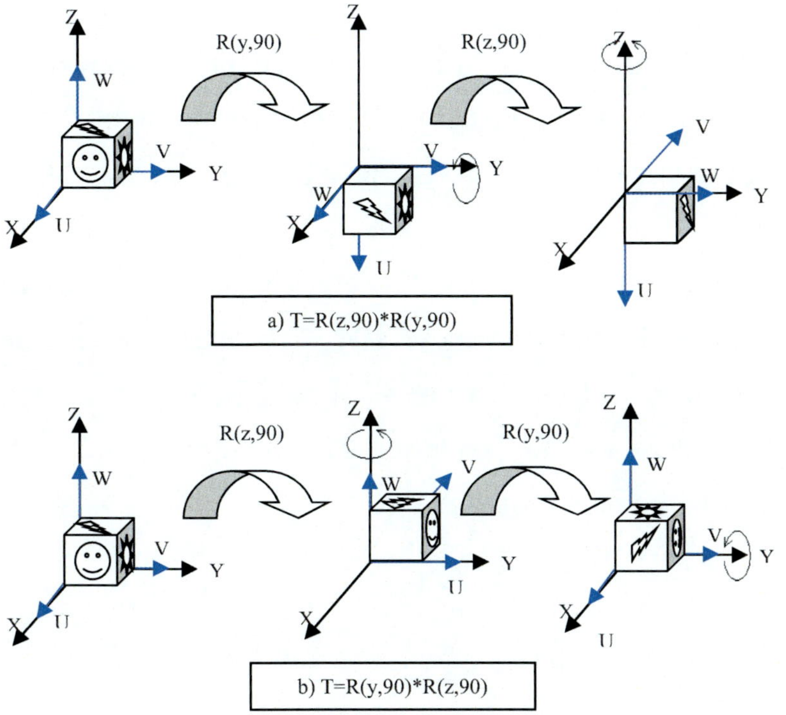

Figure 2.9 Example of Absolute Transformations.

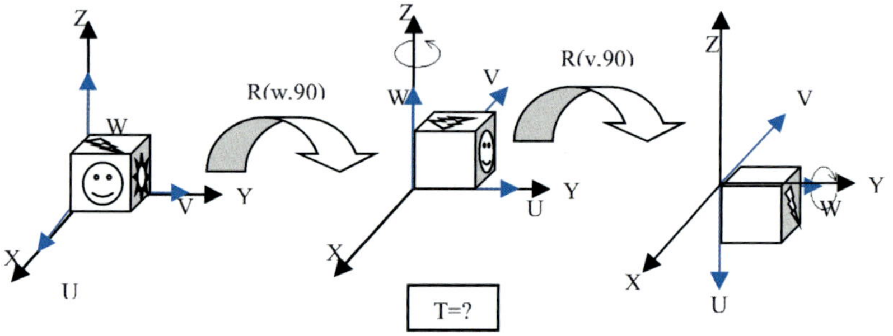

Figure 2.10 Example of Consecutive Relative Transformations

To answer the question posed in the example in Figure 2.10 will be based on a study in the plane, as was done to derive two-dimensional expression of the rotation matrix. The analysis at issue is made on the situation shown in Figure 2.11. In this case it is part of a condition in which the three frames, X_0Y_0, X_1Y_1 and X_2Y_2 are fully matched, until the situation shown in Figure 2.11; to which reaches the system after passing through the intermediate position X_1Y_1.

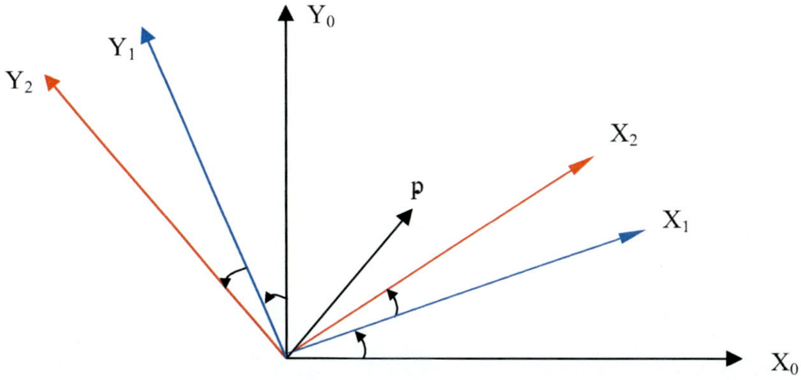

Figure 2.11 Consecutive relative rotations in the XY plane.

The point position indicated in the figure 2.11 can be written in function of their components in each reference frames.

$\overrightarrow{p_0}$: Representation of point respect to reference system X_0Y_0

$\overrightarrow{p_1}$: Representation of point respect to reference system X_1Y_1

$\overrightarrow{p_2}$: Representation of point respect to reference system X_2Y_2

Similar to as discussed for the development of the equation 2.1, these expressions can be related by rotation matrices. Here let us to introduce the notation R_1^{i-1} to refer the rotation of *i*-th system (which serves the role of mobile system), from an initial condition occupied by the previous system (fixed).

$$\overrightarrow{p_0} = R_1^0 \overrightarrow{p_1} \tag{2.3}$$
$$\overrightarrow{p_1} = R_2^1 \overrightarrow{p_2} \tag{2.4}$$

The equations 2.3 and 2.4 allow describing the consecutive changes, from X_0Y_0 status, to X_1Y_1 condition; and next, from X_1Y_1 to X_2Y_2. Similarly, is possible modelling a unique turn, from X_0Y_0 to X_2Y_2, by mean of equation 2.5.

$$\overrightarrow{p_0} = R_2^0 \overrightarrow{p_2} \tag{2.5}$$

On the other hand, the equation 2.5 has an equivalent that is easy to obtain replacing the vector $\overrightarrow{p_1}$ of equation 2.4, into the equation 2.3. The result is the equation 2.6.

$$\overrightarrow{p_0} = R_1^0 R_2^1 \overrightarrow{p_2} \tag{2.6}$$

Finally, from the equations 2.5 and 2.6 it follows that:

$$R_2^0 = R_1^0 R_2^1 \tag{2.7}$$

The equation 2.7 is the key to answer the presented question in the example of figure 2.10. Thanks to this equation the next conclusion is clear: when a rotation is realized after the condition reached with a previous rotation (**relative transformations**), the Rotation Matrices must be **post-multiplied** instead of pre-multiplied, as was required in the case of absolute transformations. Thus, the right answer to the question in the figure 2.10 is:

$$T=R(w,90)*R(v,90)$$

2.5. EULER ANGLES

As already seen, it is possible to model the changes in the orientation of the solid by combinations of rotations. RPY angles represent only one of these possibilities. In general all combinations of three consecutive rotations are known as Euler angles.

There are 6 permutations to consecutive absolute transformations: XYZ (RPY angles), XZY, YXZ, YZX, ZXY and ZYX. Another 6 sequences are added because is possible to obtain identical orientations to resultants by one rotation around an only axis, by mean of consecutive three rotations where the axis of single rotation is not included. E.g. (figure 2.12) a turn of 90º around Z axis, followed of other of 90º around X axis; and finally another of -90º, again around Z axis, will result in the same orientation than can be obtained by an unique turn of -90º around Y axis.

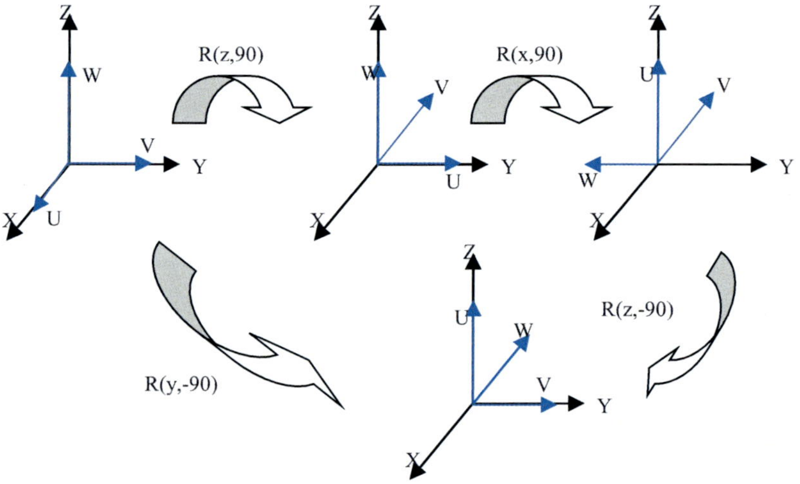

Figure 2.12 Absolute transformations. Euler angles ZXZ

The 6 sequences corresponding to this procedure are: XYX, XZX, YXY, YZY, ZXZ and ZYZ.

At last, all previous sequences can be executed as relative transformations, so that they totalize 24, commonly known as Euler angles.

2.6. HOMOGENEOUS TRANSFORMATIONS

The joint representation of the position and orientation of a solid in the space is known as **Location**. The modelling of Location can be achieved through a coordinate system called Homogeneous. It was an invention of the german mathematician August Ferdinand Möbius, which allows represent an n-dimensional space with n+1 coordinates.

The representation of vectors in Homogeneous space can be assumed as the addition of an extra coordinate, called **scale factor**, which must multiply all their components.

$$p = a\hat{i} + b\hat{j} + c\hat{k}$$

$$ph = \begin{bmatrix} aw \\ bw \\ cw \\ w \end{bmatrix}$$; w is the scale factor. It must be integer.

The value of the scale factor is arbitrary, so it is common practice to choose it equal to "1", in favor of simplicity.

Two particularities worthy of mention on the vectors in the homogeneous space: [0,0,0,0] is not defined; and those with the form [a,b,c,0] correspond to infinite longitude vectors, which they are interpreted as directions.

For matrices, the representation in Homogeneous space is achieved by increasing the size of matrix in the unity. In Robotics this increase is intentioned to achieve the combination of transformations that describe the rotation and translation processes. The result is a new matrix called Homogeneous Transformation Matrix (T), of 4x4 size, which includes: a 3x3 size submatrix to describes the changes in Orientation (Rotation Matrix), a column vector describing changes in Position (Translation Vector), a null row vector (Perspective Vector) and the scale factor (equal to 1).

$$T = \begin{bmatrix} Rotation_{3x3} & Translation_{3x1} \\ Perspective_{1x2} & Scale_{1x1} \end{bmatrix} = \begin{bmatrix} R_{3x3} & P_{3x1} \\ 0_{1x3} & 1 \end{bmatrix}$$

For any vector representative of a point in space, the application of a Homogeneous Transformation Matrix gives the result of doing rotation and translation together; which is completely applicable to the case of industrial manipulators. Indeed, given the spatial location of an object solidly attached to a reference frame OUVW whose position respect to a fixed reference system OXYZ is represented by the *ruvw* vector, can be described a change in its location by a transformation T application to mentioned vector.

$$\begin{bmatrix} x \\ y \\ z \\ 1 \end{bmatrix} = T \begin{bmatrix} u \\ v \\ w \\ 1 \end{bmatrix}$$

Here are various Transformations of particular significance:

Pure Translation

$$T(p)=\begin{bmatrix}1 & 0 & 0 & px\\ 0 & 1 & 0 & py\\ 0 & 0 & 1 & pz\\ 0 & 0 & 0 & 1\end{bmatrix} \tag{2.8}$$

Rotation around the X axis, at α angle

$$T(x,\alpha)=\begin{bmatrix}1 & 0 & 0 & 0\\ 0 & C\alpha & -S\alpha & 0\\ 0 & S\alpha & C\alpha & 0\\ 0 & 0 & 0 & 1\end{bmatrix} \tag{2.9}$$

Rotation around the Y axis, at φ angle

$$T(y,\phi)=\begin{bmatrix}C\phi & 0 & S\phi & 0\\ 0 & 1 & 0 & 0\\ -S\phi & 0 & C\phi & 0\\ 0 & 0 & 0 & 1\end{bmatrix} \tag{2.10}$$

Rotation around the Z axis, at θ angle

$$T(z,\theta)=\begin{bmatrix}C\theta & -S\theta & 0 & 0\\ S\theta & C\theta & 0 & 0\\ 0 & 0 & 1 & 0\\ 0 & 0 & 0 & 1\end{bmatrix} \tag{2.11}$$

Rotation around the X axis, at α angle, followed of a translation

$$T((x,\alpha),p)=\begin{bmatrix}1 & 0 & 0 & px\\ 0 & C\alpha & -S\alpha & py\\ 0 & S\alpha & C\alpha & pz\\ 0 & 0 & 0 & 1\end{bmatrix} \tag{2.12}$$

Rotation around the Y axis, at φ angle, followed of a translation

$$T((y,\phi),p)=\begin{bmatrix}C\phi & 0 & S\phi & px\\ 0 & 1 & 0 & py\\ -S\phi & 0 & C\phi & pz\\ 0 & 0 & 0 & 1\end{bmatrix} \tag{2.13}$$

Rotation around the Z axis, at θ angle, followed of a translation

$$T((z,\theta),p)=\begin{bmatrix}C\theta & -S\theta & 0 & px\\ S\theta & C\theta & 0 & py\\ 0 & 0 & 1 & pz\\ 0 & 0 & 0 & 1\end{bmatrix} \tag{2.14}$$

Translation followed of a rotation around the X axis

$$T(p,(x,\alpha)) = \begin{bmatrix} 1 & 0 & 0 & px \\ 0 & C\alpha & -S\alpha & pyC\alpha - pzS\alpha \\ 0 & S\alpha & C\alpha & pyS\alpha + pzC\alpha \\ 0 & 0 & 0 & 1 \end{bmatrix} \quad (2.15)$$

Translation followed of a rotation around the Y axis

$$T(p,(y,\phi)) = \begin{bmatrix} C\phi & 0 & S\phi & pxC\phi + pzS\phi \\ 0 & 1 & 0 & py \\ -S\phi & 0 & C\phi & pzC\phi - pxS\phi \\ 0 & 0 & 0 & 1 \end{bmatrix} \quad (2.16)$$

Translation followed of a rotation around the Z axis

$$T(p,(z,\theta)) = \begin{bmatrix} C\theta & -S\theta & 0 & pxC\theta - pyS\theta \\ S\theta & C\theta & 0 & pxS\theta + pyC\theta \\ 0 & 0 & 1 & pz \\ 0 & 0 & 0 & 1 \end{bmatrix} \quad (2.17)$$

The equations from 2.8 to 2.17 should be analyzed for better understanding. E.g. the equation 2.8 is distinguished because its rotation submatrix is the identity matrix, what is expected for a null rotation.

The equations 2.9, 2.10 and 2.11 that corresponding to pure rotations, have a null vector in their last column, which occurs in accordance with nonexistent translation. On the other hand, the rotation submatrices are exactly that shown in the figure 2.6, in every case.

The equations 2.12, 2.13 and 2.14 are obtained by multiplying of translation and the corresponding rotation matrices. Inside this group the submatrix of indicated rotation has a column vector attached, which is the translation vector.

Finally, the equations 2.15, 2.16 and 2.17 contain translations followed by rotations. Again is included the corresponding rotation matrix, but the vector of the fourth column is more complex than the translation vector. This result arises of multiplying (pre-multiplying) the translation transformation by the corresponding rotation transformation. The next operation illustrates the aforementioned multiplication and consists of multiplying the equations 2.11 and 2.8, yielding the equation 2.17 as result.

$$T(z,\theta)*T(p)=\begin{bmatrix} C\theta & -S\theta & 0 & 0 \\ S\theta & C\theta & 0 & 0 \\ 0 & 0 & 1 & 0 \\ 0 & 0 & 0 & 1 \end{bmatrix}\begin{bmatrix} 1 & 0 & 0 & px \\ 0 & 1 & 0 & py \\ 0 & 0 & 1 & pz \\ 0 & 0 & 0 & 1 \end{bmatrix}$$

$$=\begin{bmatrix} C\theta & -S\theta & 0 & pxC\theta - pyS\theta \\ S\theta & C\theta & 0 & pxS\theta + pyC\theta \\ 0 & 0 & 1 & pz \\ 0 & 0 & 0 & 1 \end{bmatrix}$$

2.6.1. Using Homogeneous transformations. Examples

The matrices in figure 2.6 and the equations 2.8 - 2.17 are defined respect to the axes of a fixed reference frame. Therefore, they are absolute transformations and their use to the determination of more complex transformations should be done by pre-multiplication.

In each of the following exercises are considered points attached to a mobile frame OUVW, with *uvw* coordinates. The systems initially are in full matching with a fixed frame OXYZ, and the final positions of the points in the fixed coordinate system are determined after applying some rotations and translations.

Example 1. r (u,v,w) = (2,3,0). Operations: -90° Rotation around Y axis; Translation with a vector *p* = (-4,0,-2).

The immediate solution is the application of the Homogeneous Transformation that fits the problem statement.

$$\begin{bmatrix} x \\ y \\ z \\ 1 \end{bmatrix} = T((y,\varphi),p)\begin{bmatrix} u \\ v \\ w \\ 1 \end{bmatrix} = \begin{bmatrix} C\varphi & 0 & S\varphi & px \\ 0 & 1 & 0 & py \\ -S\varphi & 0 & C\varphi & pz \\ 0 & 0 & 0 & 1 \end{bmatrix}\begin{bmatrix} u \\ v \\ w \\ 1 \end{bmatrix}$$

$$\begin{bmatrix} x \\ y \\ z \\ 1 \end{bmatrix} = \begin{bmatrix} 0 & 0 & -1 & -4 \\ 0 & 1 & 0 & 0 \\ 1 & 0 & 0 & -2 \\ 0 & 0 & 0 & 1 \end{bmatrix}\begin{bmatrix} 2 \\ 3 \\ 0 \\ 1 \end{bmatrix} = \begin{bmatrix} -4 \\ 3 \\ 0 \\ 1 \end{bmatrix}$$

Another way to solve the problem is to apply a rotation matrix and then include translation as a sum of vectors:

$$\begin{bmatrix} x \\ y \\ z \end{bmatrix} = R(y,\varphi)\begin{bmatrix} u \\ v \\ w \end{bmatrix} + \vec{p} = \begin{bmatrix} C\varphi & 0 & S\varphi \\ 0 & 1 & 0 \\ -S\varphi & 0 & C\varphi \end{bmatrix}\begin{bmatrix} 2 \\ 3 \\ 0 \end{bmatrix} + \begin{bmatrix} -4 \\ 0 \\ -2 \end{bmatrix} =$$

$$\begin{bmatrix} 0 \\ 3 \\ 2 \end{bmatrix} + \begin{bmatrix} -4 \\ 0 \\ -2 \end{bmatrix} = \begin{bmatrix} -4 \\ 3 \\ 0 \end{bmatrix}$$

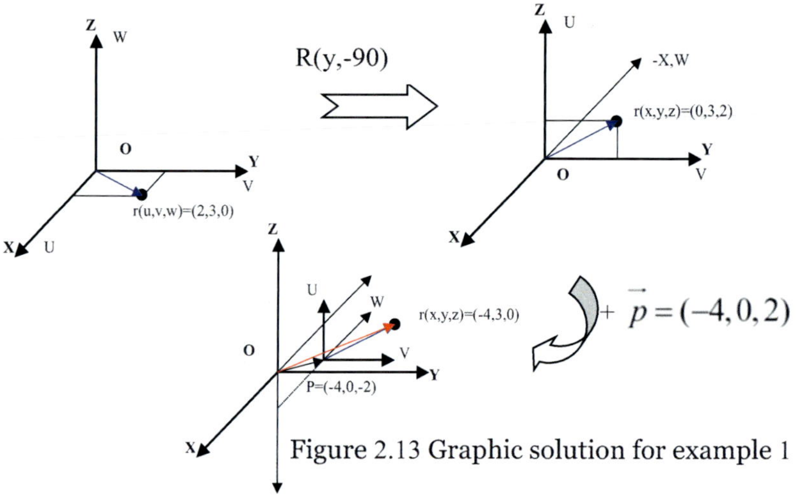

Figure 2.13 Graphic solution for example 1

Example 2. r (u,v,w) = (1,1,1). Operations: Translation with a vector *p* = (2,2,0); Rotation 45° around Z axis; Rotation 90° around Y axis.

Unlike the previous example, first it considers the Translation as a sum of vectors, and then the two rotations are made. Finally, a brief solution with Homogeneous Transformations is shown.

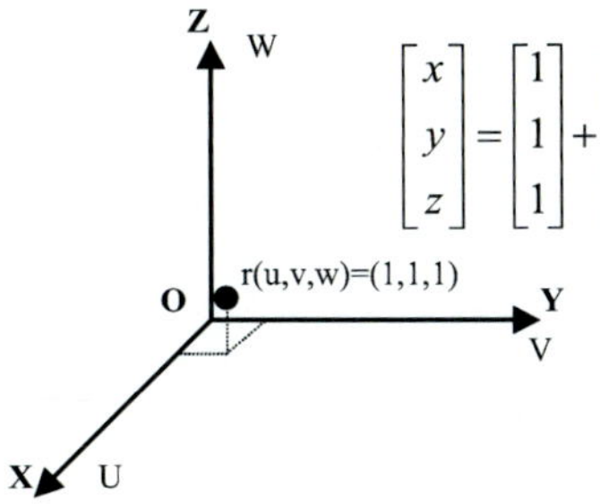

$$\begin{bmatrix} x \\ y \\ z \end{bmatrix} = \begin{bmatrix} 1 \\ 1 \\ 1 \end{bmatrix} + \begin{bmatrix} 2 \\ 2 \\ 0 \end{bmatrix} = \begin{bmatrix} 3 \\ 3 \\ 1 \end{bmatrix}$$

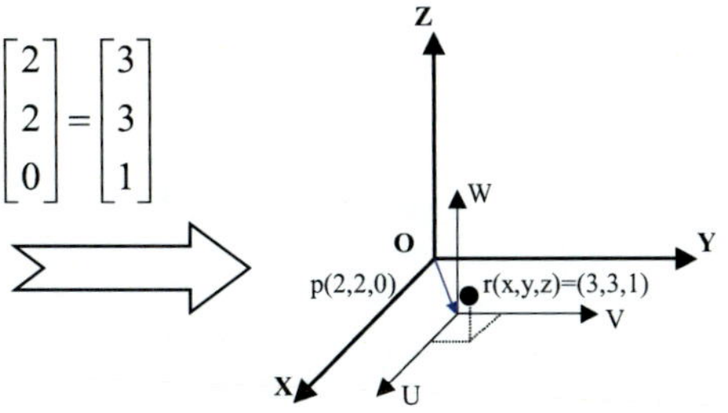

$$\begin{bmatrix} x \\ y \\ z \end{bmatrix} = R(z,\theta)\begin{bmatrix} 3 \\ 3 \\ 1 \end{bmatrix} = \begin{bmatrix} C\theta & -S\theta & 0 \\ S\theta & C\theta & 0 \\ 0 & 0 & 1 \end{bmatrix}\begin{bmatrix} 3 \\ 3 \\ 1 \end{bmatrix} = \begin{bmatrix} \sqrt{2}/2 & -\sqrt{2}/2 & 0 \\ \sqrt{2}/2 & \sqrt{2}/2 & 0 \\ 0 & 0 & 1 \end{bmatrix}\begin{bmatrix} 3 \\ 3 \\ 1 \end{bmatrix} = \begin{bmatrix} 0 \\ 3\sqrt{2} \\ 1 \end{bmatrix}$$

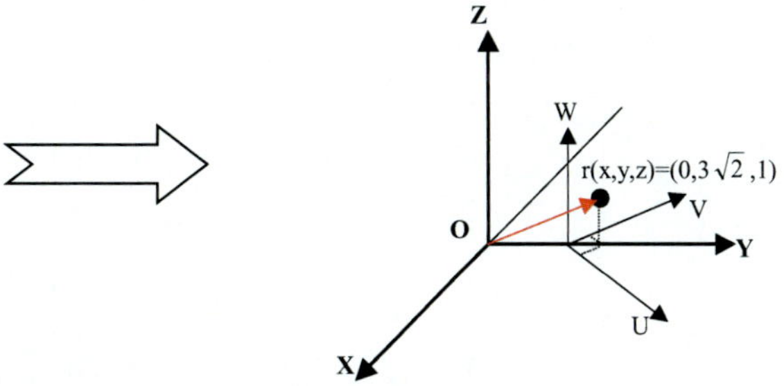

$$\begin{bmatrix} x \\ y \\ z \end{bmatrix} = R(y,\varphi)\begin{bmatrix} 0 \\ 3\sqrt{2} \\ 1 \end{bmatrix} = \begin{bmatrix} C\varphi & 0 & S\varphi \\ 0 & 1 & 0 \\ -S\varphi & 0 & C\varphi \end{bmatrix}\begin{bmatrix} 0 \\ 3\sqrt{2} \\ 1 \end{bmatrix} = \begin{bmatrix} 0 & 0 & 1 \\ 0 & 1 & 0 \\ -1 & 0 & 0 \end{bmatrix}\begin{bmatrix} 0 \\ 3\sqrt{2} \\ 1 \end{bmatrix} = \begin{bmatrix} 1 \\ 3\sqrt{2} \\ 0 \end{bmatrix}$$

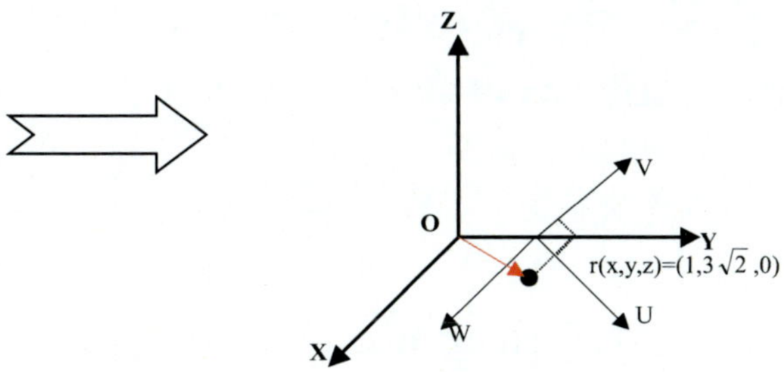

Figure 2.14 Graphic solution for example 2

A possible alternative based on Homogeneous Transformations is:

$$\begin{bmatrix} x \\ y \\ z \\ 1 \end{bmatrix} = T(y,\varphi) * T(p,(z,\theta)) \begin{bmatrix} 1 \\ 1 \\ 1 \\ 1 \end{bmatrix} =$$

$$\begin{bmatrix} C\varphi & 0 & S\varphi & 0 \\ 0 & 1 & 0 & 0 \\ -S\varphi & 0 & C\varphi & 0 \\ 0 & 0 & 0 & 1 \end{bmatrix} \begin{bmatrix} C\theta & -S\theta & 0 & pxC\theta - pyS\theta \\ S\theta & C\theta & 0 & pxS\theta + pyC\theta \\ 0 & 0 & 1 & pz \\ 0 & 0 & 0 & 1 \end{bmatrix} \begin{bmatrix} 1 \\ 1 \\ 1 \\ 1 \end{bmatrix}$$

$$\begin{bmatrix} x \\ y \\ z \\ 1 \end{bmatrix} = \begin{bmatrix} 0 & 0 & 1 & 0 \\ 0 & 1 & 0 & 0 \\ -1 & 0 & 0 & 0 \\ 0 & 0 & 0 & 1 \end{bmatrix} \begin{bmatrix} \sqrt{2}/2 & -\sqrt{2}/2 & 0 & 0 \\ \sqrt{2}/2 & \sqrt{2}/2 & 0 & 2\sqrt{2} \\ 0 & 0 & 1 & 0 \\ 0 & 0 & 0 & 1 \end{bmatrix} \begin{bmatrix} 1 \\ 1 \\ 1 \\ 1 \end{bmatrix} = \begin{bmatrix} 1 \\ 3\sqrt{2} \\ 0 \\ 1 \end{bmatrix}$$

Example 3. r (u,v,w) = (5,-7,2). Operations: Translation with a vector *p1* = (-4,4,1); Rotation -30º around Z axis; Rotation 60º around Y axis; Translation with a vector *p2* = (4,-4,-1).

We will proceed with a single approach using two Homogeneous Transformations.

$$\begin{bmatrix} x \\ y \\ z \\ 1 \end{bmatrix} = T((y,60),p2)*T(p1,(z,-30))\begin{bmatrix} 5 \\ -7 \\ 2 \\ 1 \end{bmatrix}$$

$$\begin{bmatrix} x \\ y \\ z \\ 1 \end{bmatrix} = \begin{bmatrix} C60 & 0 & S60 & 4 \\ 0 & 1 & 0 & -4 \\ -S60 & 0 & C60 & -1 \\ 0 & 0 & 0 & 1 \end{bmatrix}\begin{bmatrix} C(-30) & -S(-30) & 0 & -4C(-30)-4S(-30) \\ S(-30) & C(-30) & 0 & -4S(-30)+4C(-30) \\ 0 & 0 & 1 & 1 \\ 0 & 0 & 0 & 1 \end{bmatrix}\begin{bmatrix} 5 \\ -7 \\ 2 \\ 1 \end{bmatrix}$$

$$\begin{bmatrix} x \\ y \\ z \\ 1 \end{bmatrix} = \begin{bmatrix} 1/2 & 0 & \sqrt{3}/2 & 4 \\ 0 & 1 & 0 & -4 \\ -\sqrt{3}/2 & 0 & 1/2 & -1 \\ 0 & 0 & 0 & 1 \end{bmatrix}\begin{bmatrix} \sqrt{3}/2 & 1/2 & 0 & -4\sqrt{3}/2+4/2 \\ -1/2 & \sqrt{3}/2 & 0 & 4/2+4\sqrt{3}/2 \\ 0 & 0 & 1 & 1 \\ 0 & 0 & 0 & 1 \end{bmatrix}\begin{bmatrix} 5 \\ -7 \\ 2 \\ 1 \end{bmatrix}$$

$$\begin{bmatrix} x \\ y \\ z \\ 1 \end{bmatrix} = \begin{bmatrix} 1/2 & 0 & \sqrt{3}/2 & 4 \\ 0 & 1 & 0 & -4 \\ -\sqrt{3}/2 & 0 & 1/2 & -1 \\ 0 & 0 & 0 & 1 \end{bmatrix}\begin{bmatrix} (\sqrt{3}-3)/2 \\ (-1-3\sqrt{3})/2 \\ 3 \\ 1 \end{bmatrix}$$

$$\begin{bmatrix} x \\ y \\ z \\ 1 \end{bmatrix} = \begin{bmatrix} (7\sqrt{3}+13)/4 \\ -(3\sqrt{3}+9)/2 \\ (3\sqrt{3}-1)/4 \\ 1 \end{bmatrix}$$

2.6.2. Use of Homogeneous Transformations by means of software. Examples.

The use of equations 2.8 – 2.17 is easier with software tools. It is exemplified in this book with several scripts developed to run in Octave 4.0 or similar software. Octave is freely redistributable software under the terms of the GNU General Public License (GPL), which is downloadable in the site gnu.org.

In the following examples, the scripts of used functions are included in the appendix A, at end of the book, and you can see their descriptions in the table 2.1.

Table 2.1 Functions to apply the Homogeneous Transformations in Octave 4.0

Command	Description
Rotx (alpha)	Returns the Homogeneous transformation of rotation around X axis, an angle alpha.
Roty (phi)	Returns the Homogeneous transformation of rotation around Y axis, an angle phi.
Rotz (theta)	Returns the Homogeneous transformation of rotation around Z axis, an angle theta.
Rotxtrans (alpha,px,py,pz)	Returns the Homogeneous transformation of rotation around X axis, an angle alpha, followed by translation given in a vector p.
Rotytrans (phi,px,py,pz)	Returns the Homogeneous transformation of rotation around Y axis, an angle phi, followed by translation given in a vector p.
Rotztrans (theta,px,py,pz)	Returns the Homogeneous transformation of rotation around Z axis, an angle theta, followed by translation given in a vector p.
Transrotx (px,py,pz,alpha)	Returns the Homogeneous Transformation of translation given in a vector p, followed by rotation around X axis, an angle alpha.
Transroty (px,py,pz,phi)	Returns the Homogeneous Transformation of translation given in a vector p, followed by rotation around Y axis, an angle phi.
Transrotz (px,py,pz,theta)	Returns the Homogeneous Transformation of translation given in a vector p, followed by rotation around Z axis, an angle theta.

Example 4. r (u,v,w) = (2,1,3). Operations: Rotation 90º around X axis; Translation with a vector *p1* = (1,-1,1); Rotation -90º around Y axis; Translation with a vector *p2* = (2,-2,2).

It is possible calculate partial results using one Homogeneous Transformation by each operation. The final result will be produced by the last calculate.

```
>> ruvw = [2;1;3;1]
ruvw =

   2
   1
   3
   1

>> T1 = rotx(90)
T1 =

   1.00000    0.00000    0.00000    0.00000
   0.00000    0.00000   -1.00000    0.00000
   0.00000    1.00000    0.00000    0.00000
   0.00000    0.00000    0.00000    1.00000

>> T2 = rotxtrans(0,1,-1,1)
T2 =

   1   0   0   1
   0   1  -0  -1
   0   0   1   1
   0   0   0   1

>> T3 = roty(-90)
```

```
T3 =

   0.00000   0.00000  -1.00000   0.00000
   0.00000   1.00000   0.00000   0.00000
   1.00000   0.00000   0.00000   0.00000
   0.00000   0.00000   0.00000   1.00000

>> T4 = rotztrans(0,2,-2,2)
T4 =

   1  -0   0   2
   0   1   0  -2
   0   0   1   2
   0   0   0   1
```

It is important to notice that T2 and T4 are pure translations. They have been calculated with functions of rotations followed by translations but, and this is the essential, the angle is zero, in both cases.

Next, the rest of procedure:

```
>> pp1 = T1*ruvw
pp1 =
   2.0000
  -3.0000
   1.0000
   1.0000

>> pp2 = T2*pp1
pp2 =

   3
  -4
   2
   1
```

```
>> pp3 = T3*pp2
pp3 =

   -2.0000
   -4.0000
    3.0000
    1.0000

>> pxyz = T4*pp3
pxyz =

   2.2204e-016
   -6.0000e+000
   5.0000e+000
   1.0000e+000
```

Several comments: first, all vector have four entries, because they belong to the Homogeneous Space, where the last entry is the scale factor. Second, *pp1*, *pp2* and *pp3* are the partial results above announced. Third, the vector *pxyz* is strangely expressed and the right interpretation about its entries is (0,-6,5). The last affirmation is verifiable doing a simple null operation with *pxyz* like the following:

```
pxyz + ruvw - ruvw
ans =

    0
   -6
    5
    1
```

Finally, to this example, two alternative scripts are then copied. You can check that the produce the same result.

```
pxyz    =    rotytrans(-90,2,-2,2)*rotxtrans(90,1,-
1,1)*ruvw

pxyz    =    transrotx(2,-2,2,0)*transroty(1,-1,1,-
90)*rotx(90)*ruvw
```

Example 5. r (u,v,w) = (5,-7,2). Operations: Translation with a vector *p1* = (-4,4,1); Rotation -30º around Z axis; Rotation 60º around Y axis; Translation with a vector *p2* = (4,-4,-1).

This is the same exercise 3, but now will be solved using the Functions described above. The following line corresponds to the next script:

$$\begin{bmatrix} x \\ y \\ z \\ 1 \end{bmatrix} = T((y,60),p2)*T(p1,(z,-30))\begin{bmatrix} 5 \\ -7 \\ 2 \\ 1 \end{bmatrix}$$

```
ruvw = [5;-7;2;1]
T1 = transrotz(-4,4,1,-30)
T2 = rotytrans(60,4,-4,-1)
pxyz = T2*T1*ruvw
```

The execution of this script produces the following results:

```
>> ruvw = [5;-7;2;1]
ruvw =

    5
   -7
    2
    1

>> T1 = transrotz(-4,4,1,-30)
T1 =

    0.86603    0.50000    0.00000   -1.46410
   -0.50000    0.86603    0.00000    5.46410
    0.00000    0.00000    1.00000    1.00000
    0.00000    0.00000    0.00000    1.00000

>> T2 = rotytrans(60,4,-4,-1)
T2 =

    0.50000    0.00000    0.86603    4.00000
    0.00000    1.00000    0.00000   -4.00000
   -0.86603    0.00000    0.50000   -1.00000
    0.00000    0.00000    0.00000    1.00000

>> pxyz = T2*T1*ruvw
pxyz =

    6.2811
   -7.0981
    1.0490
    1.0000
```

CHAPTER 3

FORWARD KINEMATICS

"There is no way to peace; the peace is the way".

Mohandas Karamchand Gandhi

3.1. DEFINITION OF FORWARD KNEMATIC PROBLEM

The dilemma of determining the final position and orientation of end effector of a Robot Manipulator in relation to a fixed reference system, based on the prior knowledge of their joints and geometric parameters of its links, is called Forward Kinematic problem. That is, given an initial location of the end effector of a robot, the goal is calculating the new location, after moving their joints.

An articulated link can executes a movement respect to the previous link which may be described with a Homogeneous Transformation. For this, is necessary to establish a reference system joined at the end of each link.

The nomenclature used herein to discuss the relative motion between consecutive links uses the letter "A" to identify the above referred Transformation, assigning it the name of **A Matrix** or **Transition homogeneous matrix**. Two indices are used to refer to the related links straight through the "A" matrix: one at the upper to identify the link that works like the base of movement and another in the lower side corresponding to the mobile link. E.g. A_1^0 describes the Location of the reference system associated with the link 1 respect to the reference system associated with the link 0 (Base of the robot); A_2^1, 2 relative to 1 and A_i^{i-1}, ith relative to (i-1) th.

To a robot with six joints and seven links (Base included), the Location of end effector relative to the Base is described through a unique Matrix of Transformation called "T", which is calculated with post-multiplications of six Transition Homogeneous Matrices (Eq. 3.1).

$$T = A_1^0 * A_2^1 * A_3^2 * A_4^3 * A_5^4 * A_6^5 \tag{3.1}$$

The characterization of a Robot Manipulator through the relationship between the elements and joints is called Kinematic Chain (Richard Paul, 1981). This procedure identifies quantities and reference frames for defining the relative movement between consecutive links (**generalized coordinates**) and to generate a mathematic model for to calculate the Location of end effector relative to a fixed reference system. The method is known as Denavit-Hartenberg Algorithm, which was proposed by Jaques Denavit and Richard Scheunemann Hartenberg in a paper published in 1955.

3.2. DENAVIT-HARTENBERG PARAMETERS

The Denavit-Hartenberg Algorithm bases its utility in a specific way to select coordinate systems and determine a set of parameters (DH parameters), which will be subject of detailed study in the following lines of this section.

In principle an Industrial Robots is a kinematic chain (succession of links and joints, prismatic or revolute) where each link is associated with two joints and can be described by two parameters: **the common normal distance (a)** to the axes of joints and a **twist angle of the link (α)** (figure 3.1).

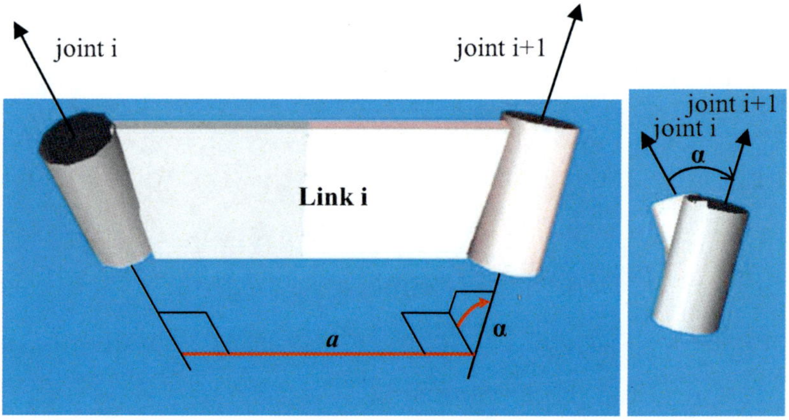

Figure 3.1 Parameters "*a*" and α

It is important to note that the angle **α** doesn't belong to the plane of frontal view in the figure 3.1, but one side. Also you should to observe how the line with the distance **"*a*"** is normal to both axes and how the planes defined by it and each axis are different.

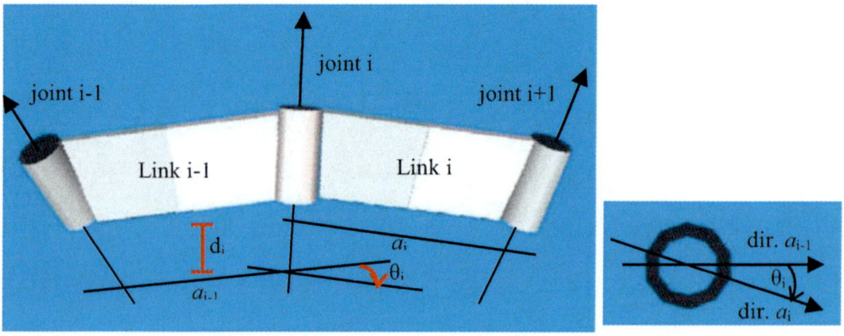

Figure 3.2 Parameters θ and d

Another two parameters can be defined from the figure 3.2, which shows two consecutive links. First let's talk about **"d"**. This is the distance between normal lines, a_{i-1} and a_i, in the direction of axis of joint i. Finally, the parameter **θ** is the angular distance from the line a_{i-1} to the line a_i.

In short:

- The DH parameters are four: θ, d, *a* and α.
- The parameters *a* and α are unique characteristics of each link; and they are always constant.
- The parameters θ and d describe the relative movement between consecutive links.
- In a revolute joint, θ is variable and d is constant.
- In a prismatic joint, d is variable and θ is constant.

3.2.1 Selecting the axes.

The origin of the reference system associated with the link "i" is located at end of itself (in coincidence with the joint "i+1"), specifically at the intersection of normal line "a_i" and the axis of the end joint of the link.

The Z axis of the link 'i' is chosen to coincide with the axis of the end joint (i+1), while the X axis is in the direction of the normal 'a_i'. The axis Y_i satisfies the product $Z_i=X_i x Y_i$.

The relationship of the DH parameters and indices of the axes can be confusing. If the reader feels a victim of this confusion must remember the following:

- The axes of an element are selected at the end, not the beginning (figure 3.3).
- The parameters "a" and α describe the element; and therefore are directly related to its system of axes, suffice it to say "mobile system".
- The parameters θ and "d" describe the relationship of this link with the previous one and therefore they are found at the initial joint and in relation to the preceding axes (with subscript 'i-1').

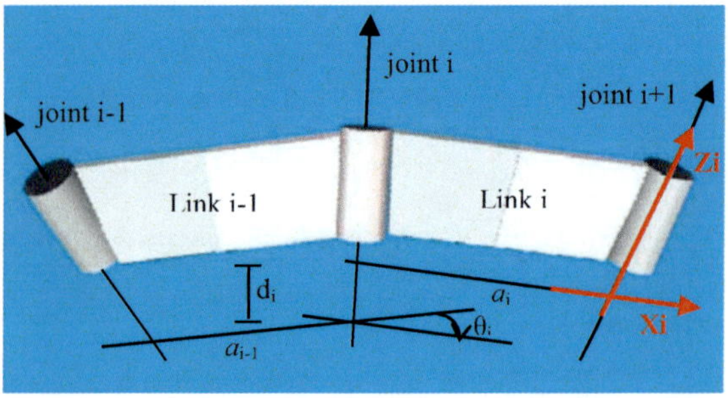

Figure 3.3 Colocation of the axes of "S_i"

3.3. TRANSITION HOMOGENEOUS MATRICES

As mentioned in a previous paragraph, a Transition homogeneous matrix describes the relative motion between two links of a kinematic chain. This description allows us to relate the reference system of element 'i' with the element 'i-1'. However, the location of a point of the element 'i' (associated with a system taken as a mobile) can be defined relative to the previous element system (i-1), taken as a fixed system, by a homogeneous transformation that considers four events:

1. Rotation around axis Z_{i-1} an angle θ_i (figure 3.4).
2. Translation along the axis Z_{i-1}, a distance d_i (figure 3.5).
3. Translation along the axis X_i, a distance a_i (figure 3.6).
4. Rotation around axis X_i, an angle α_i (figure 3.7).

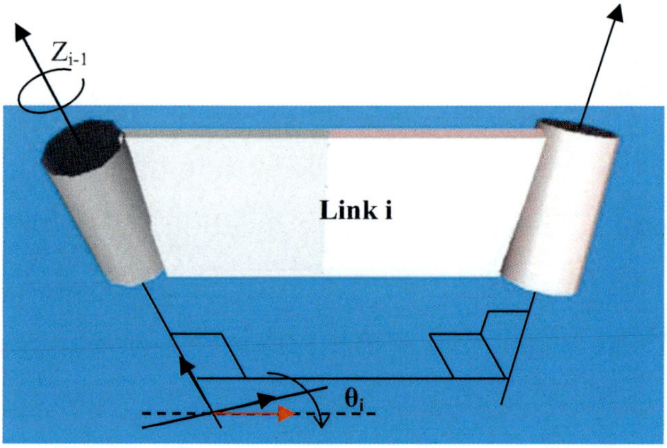

Figure 3.4 Rotation around axis Z_{i-1} an angle θ

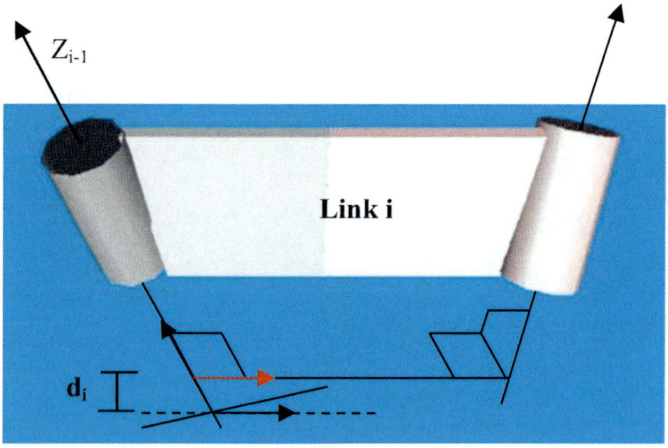

Figure 3.5 Translation along the axis Z_{i-1} a distance d

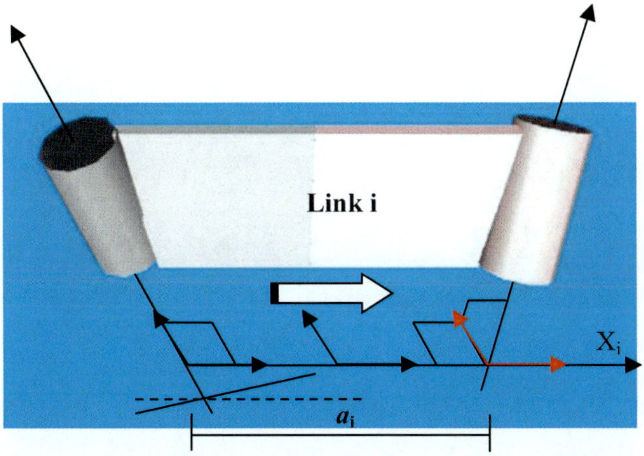

Figure 3.6 Translation along the axis X_i a distance "*a*"

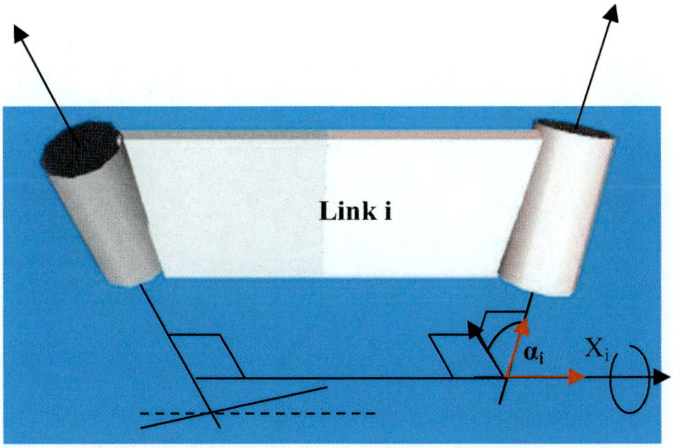

Figure 3.7 Rotation around axis X_i an angle α

As shown in the figures 3.4, 3.5, 3.6 and 3.7, these four movements let out a system of axes from a fully matching condition with a reference system ("fixed") located in the joint "i-1", to a final situation in which coincides with the joint "i". Since each movement is performed from the condition reached with the previous movement, the explicative model on entire procedure is obtained with post-multiplying of Transformations (because they are relative Transformations, see section 2.4), as it's shown below:

$$A_i^{i-1} = T(z,\theta_i)T(0,0,a_i)T(d_i,0,0)T(x,\alpha_i)$$

$$A_i^{i-1} = \begin{bmatrix} C\theta_i & -S\theta_i & 0 & 0 \\ S\theta_i & C\theta_i & 0 & 0 \\ 0 & 0 & 1 & 0 \\ 0 & 0 & 0 & 1 \end{bmatrix} \begin{bmatrix} 1 & 0 & 0 & 0 \\ 0 & 1 & 0 & 0 \\ 0 & 0 & 1 & d_i \\ 0 & 0 & 0 & 1 \end{bmatrix} \begin{bmatrix} 1 & 0 & 0 & a_i \\ 0 & 1 & 0 & 0 \\ 0 & 0 & 1 & 0 \\ 0 & 0 & 0 & 1 \end{bmatrix} \begin{bmatrix} 1 & 0 & 0 & 0 \\ 0 & C\alpha_i & -S\alpha_i & 0 \\ 0 & S\alpha_i & C\alpha_i & 0 \\ 0 & 0 & 0 & 1 \end{bmatrix}$$

$$A_i^{i-1} = \begin{bmatrix} C\theta_i & -C\alpha_i S\theta_i & S\alpha_i S\theta_i & aC\theta_i \\ S\theta_i & C\alpha_i C\theta_i & -S\alpha_i C\theta_i & aS\theta_i \\ 0 & S\alpha_i & C\alpha_i & d_i \\ 0 & 0 & 0 & 1 \end{bmatrix} \qquad (3.2)$$

3.4. DENAVIT-HARTENBERG ALGORITHM

J. Denavit and R. Hartenberg did much more than define the parameters with their names; too they established a set of instances and conventions about the axes systems associated to the links of a kinematic chain, which allow systematize the analysis of manipulator robots. Next, they are explained:

1. The links are enumerated in ascending order, beginning from the Base (number 0) to the end effector.

2. The joints are enumerated in ascending order too, assigning the number 1 to the next from the Base (figure 3.8).

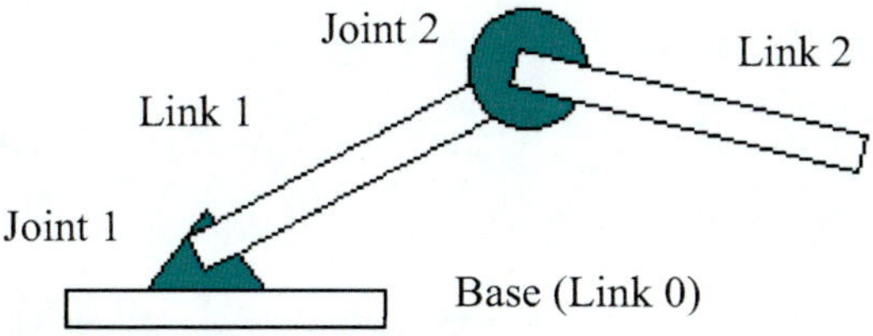

Figure 3.8 Enumeration of links and joints

3. Locate the axes of joints: in the revolute joints the axis corresponds to a turn capacity; while in the prismatic joints a displacement capacity is implemented.

4. Define the Z axis (figure 3.9). Beginning from the Base, each Z_i will be located at the joint axis corresponding to the end of the link (joint i+1).

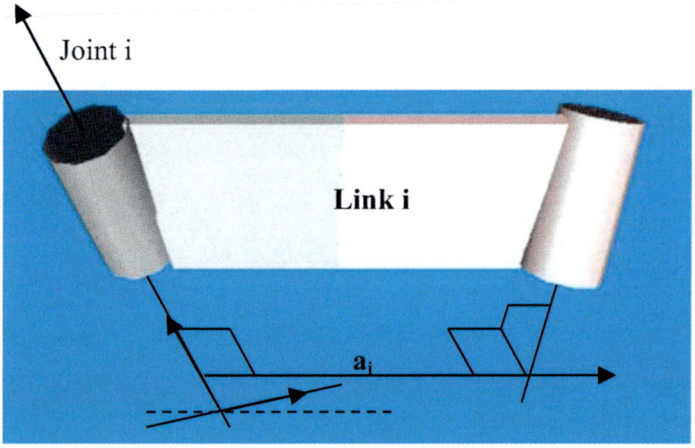

Figure 3.9 Definition of the axes according to DH conventions

5. The origin of coordinate system of the Base (S_0) may be arbitrarily chosen at any point of Z_0 axis. It will be also arbitrary the choosing of the direction of X_0 axis, but not the direction of Y_0, which is determined by rules of vector product, such that $Z_0=X_0 \times Y_0$.

6. The origin of each system 'S_i' is determined by the intersection of normal line a_i and Z_i. The X_i axis is the direction of the normal line a_i (figure 3.9) and the corresponding Y_i satisfies the rule of all dextrorotatory, $Z_i=X_i \times Y_i$.

7. The system 'S_n' is located at the end effector of the robot, with origin in the TCP and establishing coincidence between the directions of Z_n and Z_{n-1}. El sistema 'S_n' se ubicará en el elemento terminal del Robot y el eje Z_n conservará la dirección de Z_{n-1}.

8. $\boldsymbol{\theta_i}$ is defined as the angle around Z_{i-1} from X_{i-1} to X_i.

9. $\mathbf{d_i}$ is the offset along Z_{i-1} from X_{i-1} to X_i.

10. $\boldsymbol{a_i}$ is the displacement along X_i, from the origin of S_{i-1} of the origin of S_i.

11. $\boldsymbol{\alpha_i}$ is the angle around X_i, from Z_{i-1} to Z_i.

The last four are the classic definitions of the Denavit Hartenberg parameters, used to make a table with a number of rows equal than the quantity of joints (called Table of DH Parameters). About them is remarkable the use of the right-hand rule to determine the **signs of the angles θ and α** (figures 3.10 and 3.11). The sign of θ_i is positive (+) when the right thumb is pointing to the positive direction of Z_{i-1} while the other four fingers are simulating the gyre of X_{i-1} to X_i.

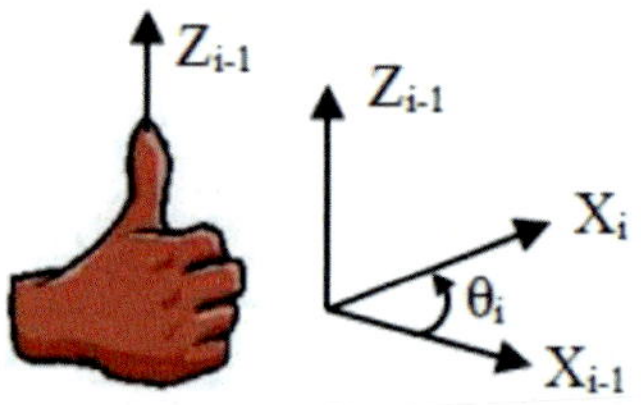

Figure 3.10 Right Hand Rule. Case: θ_i

The same is observable to the case of α: The sign of α_i is positive (+) when the right thumb is pointing to the positive direction of X_{i1} while the other four fingers are simulating the gyre of Z_{i-1} to Z_i.

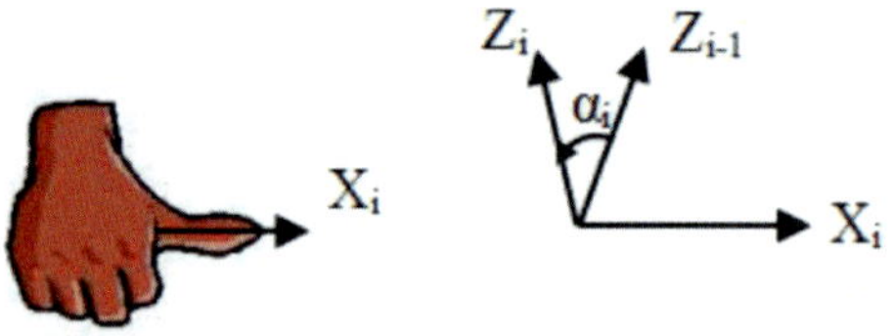

Figure 3.11 Right Hand Rule. Case: α_i

12. Determine the expressions of Transition Homogeneous Matrices (A_i^{i-1}). There will be one

Transition homogeneous matrix for each joint of the robot.

13. Determine the Transformation Matrix of the robot (T) with the multiplying of all Transition Homogeneous Matrices:

$$T = A_1^0 * A_2^1 * \ldots * A_n^{n-1}$$

Depending on the type of joint used, each Transition Matrix Homogeneous has as variable the angle θ (revolute joint) or the distance d (prismatic joint). These two are called articulation parameter (q) and its conscious variation allows controlling the Location of the end effector. A simple Boolean equation discriminates whether the parameter "q" is θ or d, by a logical variable called parameter Vereshagin (σ), which is zero if the joint is rotary or 1 if it is prismatic.

$$q = \overline{\sigma}\theta + \sigma d \tag{3.3}$$

In synthesis, the solution of Forward Kinematic problem is the Transformation Matrix of Robot (T), which is determined with the Transition Homogeneous Matrices based on the DH parameters associated with each joint (pair of consecutive links). With these tools and the control of the joints parameters is achieved positioning and orienting the end effector where it is desired.

CHAPTER 4

INVERSE KINEMATICS AND DIFFERENTIAL KINEMATICS

Work of thought is like drilling a well: the water is cloudy at first, then, it clarifies.

Chinese Proverb

4.1 INVERSE KINEMATICS

4.1.1. Definition of the Inverse Kinematic Problem

Another fundamental problem in robotics is to determine how to handle the parameters of a robot so that its tool (end effector) reaches a desired position and orientation. Determine the set of values of the parameters θ d of a robot, satisfying the need to orient and position its tool on a specific point, is the dilemma known as Inverse Kinematic Problem.

In Forward Kinematics the position and orientation are functions that depend on the joint parameters:

$$\vec{p} = \begin{cases} f_{1x}(q_1, q_2, ..., q_n) \\ f_{2x}(q_1, q_2, ..., q_n) \\ f_{3x}(q_1, q_2, ..., q_n) \end{cases} \quad \vec{n} = \begin{cases} f_{4x}(q_1, q_2, ..., q_n) \\ f_{5x}(q_1, q_2, ..., q_n) \\ f_{6x}(q_1, q_2, ..., q_n) \end{cases}$$

$$\vec{s} = \begin{cases} f_{7x}(q_1, q_2, ..., q_n) \\ f_{8x}(q_1, q_2, ..., q_n) \\ f_{9x}(q_1, q_2, ..., q_n) \end{cases} \quad \vec{a} = \begin{cases} f_{10x}(q_1, q_2, ..., q_n) \\ f_{11x}(q_1, q_2, ..., q_n) \\ f_{12x}(q_1, q_2, ..., q_n) \end{cases}$$

In Inverse Kinematics the joints parameters are functions that depend on the position and orientation:

$$\begin{aligned} q_1 &= g_1(\vec{p}, \vec{n}, \vec{s}, \vec{a}) \\ q_2 &= g_2(\vec{p}, \vec{n}, \vec{s}, \vec{a}) \\ &. \\ &. \\ &. \\ q_n &= g_n(\vec{p}, \vec{n}, \vec{s}, \vec{a}) \end{aligned} \tag{4.1}$$

4.1.2. Possible results of Inverse Kinematics

Faced with a task to perform, an industrial manipulator can reach a particular point in its trajectory by a specific combination of its joint parameters. It can also happen that there are several combinations to solve the problem; or perhaps there is no way can be achieved (figure 4.1).

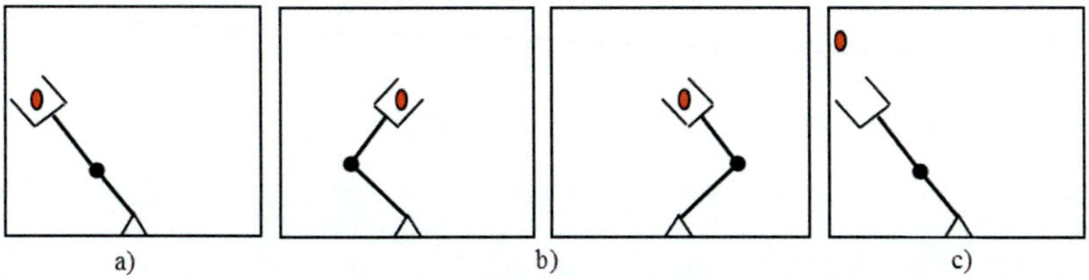

Figure 4.1 Possible results of Inverse Kinematics: a) Unique solution, b) Multiple solutions, c) No solution

In short, there are three possible events in the inverse kinematics:

- There is an unique solution
- There are multiple solutions
- There is no solution

How to tackle the problem?

To get the particular solutions of the equation 4.1, we can appeal to trigonometric studies or use analytical procedures. The first option, although always possible, requires great powers of observation and some skill as a draftsman. Using it is convenient when it comes to robots with few joints and simple configurations.

The second alternative is possible, among other procedures, from the mathematical model obtained by the Denavit-Hartenberg algorithm (Matrix T) to solve the direct kinematic problem; but this does not always have a set of equations from which we can isolating the parameters θ and d of the joints. In this case, the aim to propose a system with as many equations as unknowns (θ and d), it can be achieved by developing the pre-multiplications shown in the equation 4.2.

$$
\begin{gathered}
T = A_1^0 * A_2^1 * \ldots * A_n^{n-1} \\
\left(A_1^0\right)^{-1} * T = A_2^1 * A_3^2 * \ldots * A_n^{n-1} \\
\left(A_2^1\right)^{-1} * \left(A_1^0\right)^{-1} * T = A_3^2 * \ldots * A_n^{n-1} \\
\left(A_{n-1}^{n-2}\right)^{-1} * \ldots * \left(A_2^1\right)^{-1} * \left(A_1^0\right)^{-1} * T = A_n^{n-1}
\end{gathered}
\tag{4.2}
$$

It should be noted that the Matrix T is used under algebraic expression in this equation.

$$T = \begin{bmatrix} nx & sx & ax & px \\ ny & sy & ay & py \\ nz & sz & az & pz \\ 0 & 0 & 0 & 1 \end{bmatrix}$$

Inverse Kinematic The problem may actually be a real formidable "gargle" to a computer system that has microprocessors and most powerful software tools. This is due to the following: the matrix equation 4.2 is just the starting point for, by comparing elements in both sides of the equation, a system of linearly independent equations. Solve this system of equations is another story, since the variables to isolate usually are on the arguments transcendental functions (it is the θ angles in the Sine and Cosine functions). The following chapters illustrate how it is possible to follow this procedure.

4.2. DIFERENTIAL KINEMATICS

The subject matter of the differential kinematics is the ratio of the velocities of the joints and the velocity of the end effector. That relationship is described by a matrix called Jacobian matrix, which expresses speeds, linear and rotational of the end effector, as functions of speeds joints (q').

Express the linear velocity is easy because depends on the variation of an only vector, respect to the time, the position vector (p), but is more difficult talk about the rotational velocity based on the timing variation of the unit vectors normal, slide and approach (n, s and a). In this case use the Euler angles RPY, Roll (φ), Pitch (θ) and Yaw (ψ) to achieve the representation of the orientation, is better. Then, is possible work with a set of 6 independent variables (known as robot's operational space) to doing the study of end effector trajectories (changes of position and orientation relative to the time) by mean of their derivatives and their relationships with the joints parameters.

$$
\begin{aligned}
x &= fx(q_1, q_2, \ldots, q_n) \\
y &= fy(q_1, q_2, \ldots, q_n) \\
z &= fz(q_1, q_2, \ldots, q_n) \\
\varphi &= f\varphi(q_1, q_2, \ldots, q_n) \\
\theta &= f\theta(q_1, q_2, \ldots, q_n) \\
\psi &= f\psi(q_1, q_2, \ldots, q_n)
\end{aligned}
$$

After taking derivatives:

$$x' = \frac{\partial fx}{\partial q_1} q_1' + \frac{\partial fx}{\partial q_2} q_2' + \ldots + \frac{\partial fx}{\partial q_n} q_n'$$

$$y' = \frac{\partial fy}{\partial q_1} q_1' + \frac{\partial fy}{\partial q_2} q_2' + \ldots + \frac{\partial fy}{\partial q_n} q_n'$$

.

.

.

$$\psi' = \frac{\partial f\psi}{\partial q_1} q_1' + \frac{\partial f\psi}{\partial q_2} q_2' + \ldots + \frac{\partial f\psi}{\partial q_n} q_n'$$

The matrix relationship which produces the above expressions is the Jacobian Matrix (J). The Jacobian Matrix contains the partial derivatives of the functions that express the position and orientation of the end effector in terms of the joints parameters.

$$\begin{bmatrix} x' \\ y' \\ z' \\ \varphi' \\ \theta' \\ \psi' \end{bmatrix} = J \begin{bmatrix} q_1' \\ q_2' \\ . \\ . \\ . \\ q_n' \end{bmatrix}$$

$$J = \begin{bmatrix} \frac{\partial fx}{\partial q_1} & \frac{\partial fx}{\partial q_2} & \cdot\ \cdot\ \cdot & \frac{\partial fx}{\partial q_n} \\ \frac{\partial fy}{\partial q_1} & \frac{\partial fy}{\partial q_2} & \cdot\ \cdot\ \cdot & \frac{\partial fy}{\partial q_n} \\ \cdot & \cdot & & \cdot \\ \cdot & \cdot & & \cdot \\ \cdot & \cdot & & \cdot \\ \frac{\partial f\psi}{\partial q_1} & \frac{\partial f\psi}{\partial q_2} & \cdot\ \cdot\ \cdot & \frac{\partial f\psi}{\partial q_n} \end{bmatrix} \tag{4.3}$$

4.3.1. Inverse Jacobian Matrix

For the same reason the Jacobian matrix allows to know the speed of the end effector from the joint velocities through the inverse Jacobian matrix (J^{-1}) can be determined joint velocities based on the knowledge of the instantaneous speed of end effector.

$$\begin{bmatrix} q_1' \\ q_2' \\ \cdot \\ \cdot \\ \cdot \\ q_n' \end{bmatrix} = J^{-1} \begin{bmatrix} x' \\ y' \\ z' \\ \varphi' \\ \theta' \\ \psi' \end{bmatrix}$$

However, getting the Inverse Jacobian Matrix may be very complicated, so is relevant discuss about the possible ways for doing it.

The longer procedure is to take partial derivatives with respect to the variables defining the position and orientation, each of the functions obtained by the inverse kinematic study. Just remember the tremendous complication that exists for such equations, to want to opt for an alternative.

$$
\begin{aligned}
q_1 &= f1(x, y, z, \varphi, \theta, \psi) \\
q_2 &= f2(x, y, z, \varphi, \theta, \psi) \\
&. \\
&. \\
&. \\
q_n &= fn(x, y, z, \varphi, \theta, \psi)
\end{aligned}
$$

$$
\begin{aligned}
q_1' &= \frac{\partial f1}{\partial x}x' + \frac{\partial f1}{\partial y}y' + \ldots + \frac{\partial f1}{\partial \psi}\psi' \\
q_2' &= \frac{\partial f2}{\partial x}x' + \frac{\partial f2}{\partial y}y' + \ldots + \frac{\partial f2}{\partial \psi}\psi' \\
&. \\
&. \\
&. \\
q_n' &= \frac{\partial fn}{\partial x}x' + \frac{\partial fn}{\partial y}y' + \ldots + \frac{\partial fn}{\partial \psi}\psi'
\end{aligned}
$$

$$J^{-1} = \begin{bmatrix} \frac{\partial f1}{\partial x} & \frac{\partial f1}{\partial y} & . \; . \; . & \frac{\partial f1}{\partial \psi} \\ \frac{\partial f2}{\partial x} & \frac{\partial f2}{\partial y} & . \; . \; . & \frac{\partial f2}{\partial \psi} \\ . & . & & . \\ . & . & & . \\ . & . & & . \\ \frac{\partial fn}{\partial x} & \frac{\partial fn}{\partial y} & . \; . \; . & \frac{\partial fn}{\partial \psi} \end{bmatrix} \tag{4.4}$$

The most reasonable option is based on direct kinematic study and Jacobian Matrix, and then inverting the latter. The problem is that the Jacobian Matrix (J) is not always square and in case if it is, sometimes has a determinant (Jacobian) void; suffice it to say, no inverse.

When J is not square is due to presume to know the six variables that characterize the entire operational space (x, y, z, φ, θ, ψ) and not having six degrees of freedom (about six) in the robot manipulator studied. **Over 6 DOF** means any position and orientation can be achieved without moving a joint, so the corresponding joint velocity can be assumed as zero and work with 6x6 matrices. **Less than 6 DOF** implies that it is impossible for the end effector acquire some guidance so that one or more of the RPY angles can be excluded from cinematic study and work again with a square matrix.

If working with J numerically, it may happen that J^{-1} does not exist. In such circumstances the use of a reverse pseudo as $(J^*J^T)^{-1}$ can be a good alternative.

4.3.2. Uses of Jacobian Matrix

As already have mentioned, the evaluation of the Jacobian Matrix is used to determine the speed of translation and rotation of the end effector, but only for an instant. Indeed, all the elements of equation 4.4 are algebraic expressions dependent parameters "θ and d", which define the robot kinematics. For each angular position θ of each joint rotation and linear position "d" of the prismatic joints, equation 4.4 acquires a fully numerical expression, which can be applied to the vector of joint velocities (q ') for determining the instantaneous values of the translation speed of the TCP and variations versus time of the quantities which define the angular orientation of the tool. So that repeated evaluation of the Jacobian matrix, only contributes to the determination of curves of instantaneous speed and requires instrumentation information from accompanying the robot.

Another application of the Jacobian matrix begins with the calculation of its determinant. The determinant of the Jacobian matrix is often used to determine the values of the angles and displacements for a singularity occurs, i.e., values for which it has no inverse. These values are particularly important to define the limits of the workload of the handlers. The solution of the equation 4.5 allows calculating these values.

$$|J| = 0 \tag{4.5}$$

Applications of this equation are provided in the chapters pertaining to kinematic studies of some of the most widely used Industrial Robots.

CHAPTER 5

KINEMATICS OF CARTESIAN ROBOT

"The future starts today, not tomorrow"

Karol Józef Wojtyła

Here the simplest of Industrial Robots. To this model has been added a fourth revolute joint.

5.1. FUNCTIONAL REPRESENTATION AND COORDINATE SYSTEMS

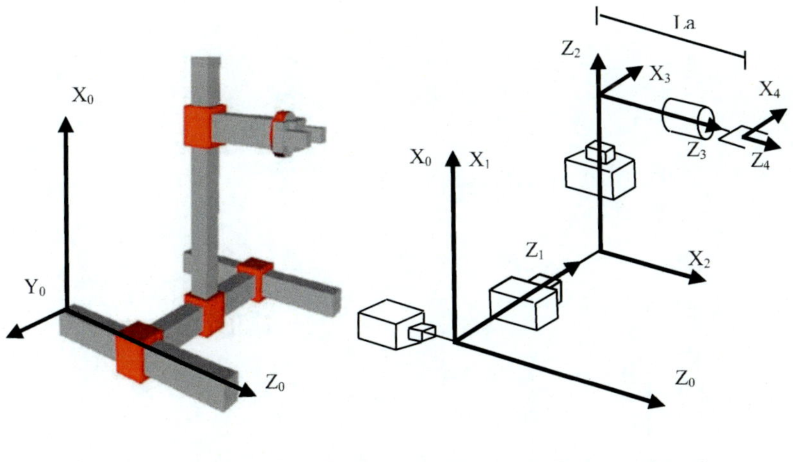

Figure 5.1 Functional representation and generalized coordinates systems of Cartesian Robot

5.2. TABLE OF DH PARAMETERS

Table 5.1 Table of DH parameters of Cartesian Robot

Joint	θ	d	a	α	q
1	0	q_1	0	90	d_1
2	90	q_2	0	90	d_2
3	90	q_3	0	90	d_3
4	q_4	La	0	0	θ_4

5.3 TRANSITION HOMOGENEOUS MATRICES

General equation:

$$^{i-1}A_i = \begin{bmatrix} C\theta & -C\alpha S\theta & S\alpha S\theta & aC\theta \\ S\theta & C\alpha C\theta & -S\alpha C\theta & aS\theta \\ 0 & S\alpha & C\alpha & d \\ 0 & 0 & 0 & 1 \end{bmatrix}$$

Particular matrices:

$$A_1^0 = \begin{bmatrix} 1 & 0 & 0 & 0 \\ 0 & 0 & -1 & 0 \\ 0 & 1 & 0 & d1 \\ 0 & 0 & 0 & 1 \end{bmatrix} \qquad A_2^1 = \begin{bmatrix} 0 & 0 & 1 & 0 \\ 1 & 0 & 0 & 0 \\ 0 & 1 & 0 & d2 \\ 0 & 0 & 0 & 1 \end{bmatrix}$$

$$A_3^2 = \begin{bmatrix} 0 & 0 & 1 & 0 \\ 1 & 0 & 0 & 0 \\ 0 & 1 & 0 & d3 \\ 0 & 0 & 0 & 1 \end{bmatrix} \qquad A_4^3 = \begin{bmatrix} C4 & -S4 & 0 & 0 \\ S4 & C4 & 0 & 0 \\ 0 & 0 & 1 & La \\ 0 & 0 & 0 & 1 \end{bmatrix}$$

5.4. TRANSFORMATION MATRIX OF THE ROBOT

$$T = A_1^0 * A_2^1 * A_3^2 * A_4^3$$

$$T = \begin{bmatrix} S4 & C4 & 0 & d3 \\ -C4 & S4 & 0 & -d2 \\ 0 & 0 & 1 & La + d1 \\ 0 & 0 & 0 & 1 \end{bmatrix}$$

The equations of position are what you might expect:

$$\boxed{\begin{aligned} px &= d_3 \\ py &= -d_2 \\ pz &= La + d_1 \end{aligned}}$$

5.5. ANALYSIS OF RESULTS

The axes of the Base reference system have been chosen according with the DH conventions. However, another natural selection of the axes is shown in the figure 5.2.

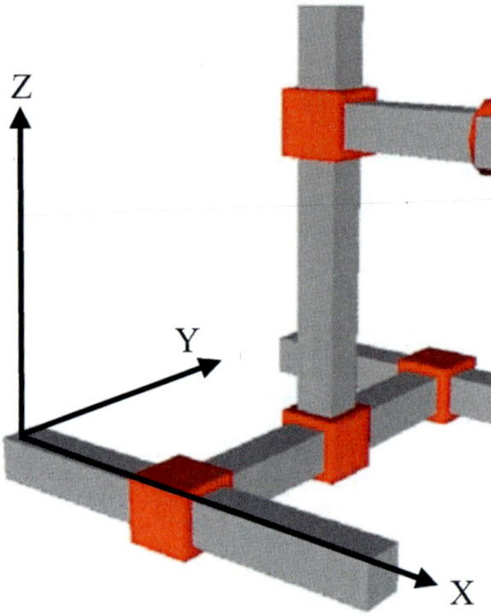

Figure 5.2 Alternative selection for the axes of Cartesian Robot

For this new system of axes a simple adaptation of the previous results produces the following expressions, which are more fitted to what you would expect for this robot.

$$pz = d_3$$

$$py = d_2$$

$$px = La + d_1$$

Respect to the vectors defining the orientation (*n*, *s* and *a*), a simple inspection of the T matrix shows the *approach* vector without projection on the axes X_0, Y_0 (figure 5.3).

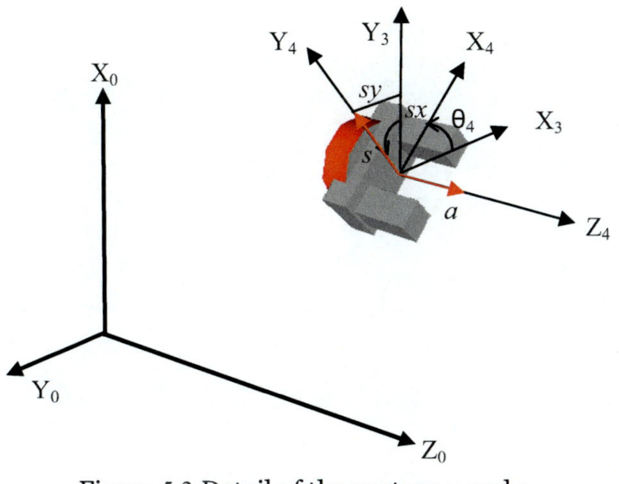

Figure 5.3 Detail of the vectors *a* and *s*

Meanwhile, an elegant relationship between the vector *s* and the angle θ_4 is easily obtained by dividing the entries (2,2) and (2,1) of the T matrix:

$$\boxed{\frac{sy}{sx} = Tg\theta_4}$$

CHAPTER 6

KINEMATICS OF CYLINDRICAL ROBOT

"Dripping water hollows out stone, not through force but through persistence."

Publio Ovidio Nasón

6.1. FUNCTIONAL REPRESENTATION AND COORDINATE SYSTEMS

The following analysis corresponds to the Cylindrical-UC Robot which was together developed by the School of Mechanic Engineering and the Laboratory of Robotics and Industrial Vision of the School of Electrical Engineering of the University of Carabobo (Venezuela). This Manipulator has a revolute joint constituting a fourth degree of freedom added after the standard RPP configuration of all cylindrical robots.

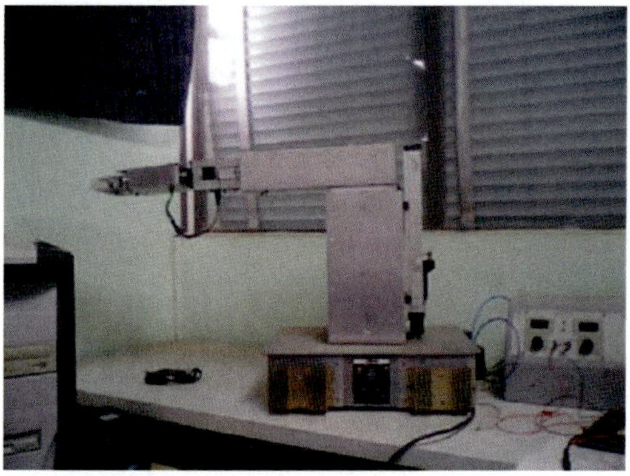

Figure 6.1 Cylindrical-UC Robot
Courtesy of University of Carabobo

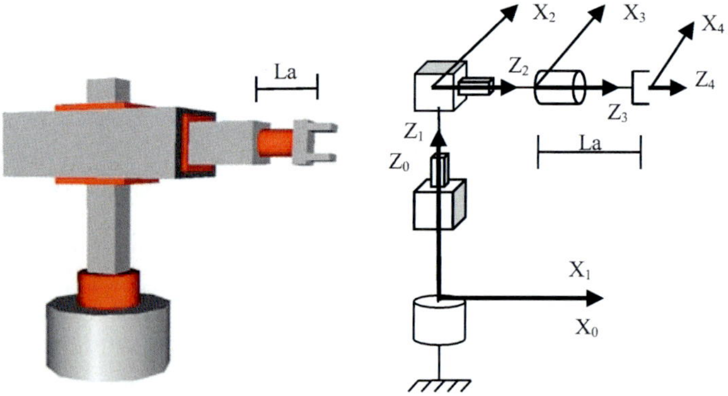

Figure 6.2 Functional representation and generalized coordinates systems of Cylindrical Robot

6.2. TABLE OF DH PARAMETERS

Table 6.1 Table of DH parameters Cylindrical Robot UC

Joint	θ	d	a	α	q
1	q_1	0	0	0	θ_1
2	90	q_2	0	90	d_2
3	0	q_3	0	0	d_3
4	q_4	La	0	0	θ_4

6.3 TRANSITION HOMOGENEOUS MATRICES

General equation:

$$^{i-1}A_i = \begin{bmatrix} C\theta & -C\alpha S\theta & S\alpha S\theta & aC\theta \\ S\theta & C\alpha C\theta & -S\alpha C\theta & aS\theta \\ 0 & S\alpha & C\alpha & d \\ 0 & 0 & 0 & 1 \end{bmatrix}$$

Particular matrices:

$$A_1^0 = \begin{bmatrix} C1 & -S1 & 0 & 0 \\ S1 & C1 & 0 & 0 \\ 0 & 0 & 1 & 0 \\ 0 & 0 & 0 & 1 \end{bmatrix} \quad A_2^1 = \begin{bmatrix} 0 & 0 & 1 & 0 \\ 1 & 0 & 0 & 0 \\ 0 & 1 & 0 & d2 \\ 0 & 0 & 0 & 1 \end{bmatrix}$$

$$A_3^2 = \begin{bmatrix} 1 & 0 & 0 & 0 \\ 0 & 1 & 0 & 0 \\ 0 & 0 & 1 & d3 \\ 0 & 0 & 0 & 1 \end{bmatrix} \quad A_4^3 = \begin{bmatrix} C4 & -S4 & 0 & 0 \\ S4 & C4 & 0 & 0 \\ 0 & 0 & 1 & La \\ 0 & 0 & 0 & 1 \end{bmatrix}$$

6.4. TRANSFORMATION MATRIX OF THE ROBOT

$$T = A_1^0 * A_2^1 * A_3^2 * A_4^3$$

$$T = \begin{bmatrix} -S1C4 & S1S4 & C1 & C1(La + d_3) \\ C1C4 & -C1S4 & S1 & S1(La + d_3) \\ S4 & C4 & 0 & d_2 \\ 0 & 0 & 0 & 1 \end{bmatrix}$$

The equations of position are:

$$\boxed{px = C1(La + d_3)}$$

$$\boxed{py = S1(La + d_3)}$$

$$\boxed{pz = d_2}$$

6.5. ANALYSIS OF RESULTS

The functional representation used to get the model (Matrix T) shows a condition of robot where $px= d_3+La$, $py=0$, $pz=d_2$. This position implies what the joints parameters one and four must be nulls ($\theta_1= \theta_4= 0^o$); and certainly, the position resulting when the above equations are evaluated with these values coincides totally with it expected.

Special comments deserve the vectors *n*, *s* and *a* defining the orientation. Is remarkable, e.g., how the matrix T includes a null projection of the *approach* on the Z_0 axis, while the value $\theta_1=0^o$ implies that $a_y = 1$, the total magnitude of the *approach* to its projection on the Y_0 axis! A similar analysis is appropriate to the vectors *slide* and *normal*, whose elements are entirely dependent on the angular values of the parameters one and four.

6.6. INVERSE KINEMATICS

Due to the configuration of the robot, the simplest way to approach the study of inverse kinematics is through a geometric analysis in the cylindrical coordinate system (Figure 6.3).

$$r = \sqrt{x^2 + y^2} \tag{6.1}$$

$$\theta = Tg^{-1}\left(y / x\right) \tag{6.2}$$

$$z = z \tag{6.3}$$

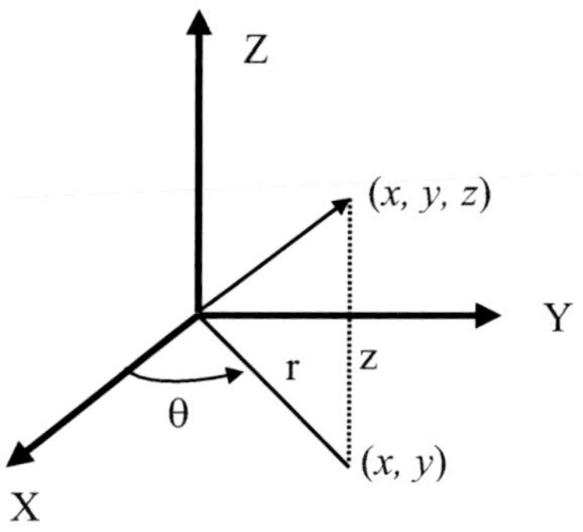

Figure 6.3. Cylindrical Coordinate System

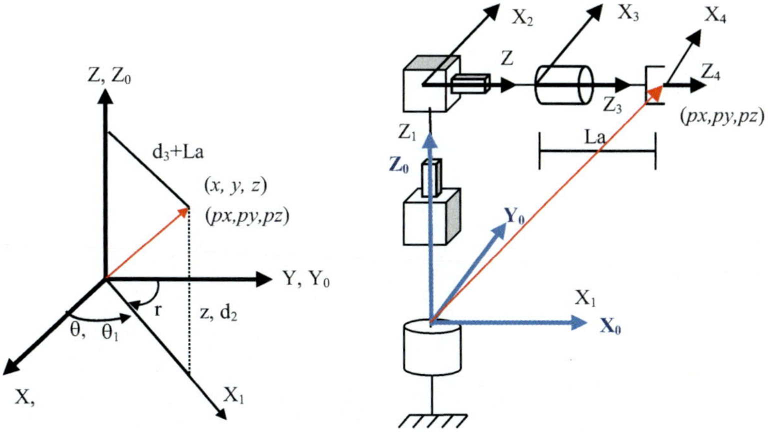

Figure 6.4 Robot model adaptations to the Cylindrical Coordinate System

As seen in Figure 6.4 the TCP position (px, py, pz) can be assimilated to any point in the Cartesian coordinate system (x, y, z) and applying the relationship between the robot parameters and variables the cylindrical coordinate system as follows:

$$r = d_3 + La$$

$$z = d_2$$

As regards the angle θ, it is easier to write directly an equation in terms of θ_1, instead of manipulating a difference of angles:

$$Tg(\theta_1) = py / px$$

With this latest result and the replacement of "*r*" it comes to obtaining the inverse kinematic model.

$$d_3 = \sqrt{px^2 + py^2} - La$$

$$\theta_1 = Tg^{-1}\left(px / py\right)$$

$$d_2 = pz$$

As regards the angle θ_4, is required analyze the relationship between the axes X_3 and X_4 from a frontal view to the *approach* vector (figure 6.5).

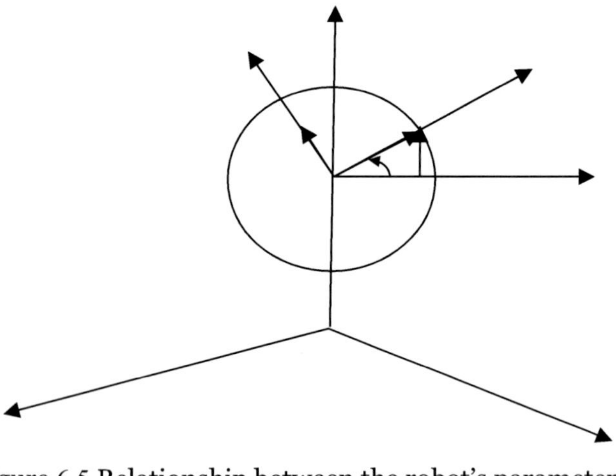

Figure 6.5 Relationship between the robot's parameters and the vectors defining the orientation

The projections of vectors *normal* and *slide* on the axes X_0 and Y_0, interchange their magnitudes depending on the angle θ_1. For this reason the following analysis only tries about the projections on the axis Z_0 (n_z and s_z), which are functions of the angle θ_4. So, from the small triangle drawn in the figure 6.5 is possible deriving the next relationship:

$$n_z = |\vec{n}| Sin(\theta_4)$$

Given that $|\vec{n}| = 1$,

$$\boxed{\theta_4 = Sin^{-1}(n_z)}$$

The alternative analytical procedure is not easy. As already said, this analytical procedure is based on the Denavit-Hartenberg algorithm and the algebraic expression of the Matrix T.

We proceed to the determination of Transition Homogeneous matrices, their inverses and the extraction of an equations system obtained from the comparison of elements within matrices.

$$T = A_1^0 * A_2^1 * A_3^2 * A_4^3$$

$$\left(A_1^0\right)^{-1} * T = A_2^1 * A_3^2 * A_4^3$$
$$\left(A_2^1\right)^{-1} * \left(A_1^0\right)^{-1} * T = A_3^2 * A_4^3$$
$$\left(A_3^2\right)^{-1} * \left(A_2^1\right)^{-1} * \left(A_1^0\right)^{-1} * T = A_4^3 \tag{6.4}$$

Next, all required factors for write the above equations.

$$A_1^0 = \begin{bmatrix} C1 & -S1 & 0 & 0 \\ S1 & C1 & 0 & 0 \\ 0 & 0 & 1 & 0 \\ 0 & 0 & 0 & 1 \end{bmatrix} \quad \left(A_1^0\right)^{-1} = \begin{bmatrix} C1 & S1 & 0 & 0 \\ -S1 & C1 & 0 & 0 \\ 0 & 0 & 1 & 0 \\ 0 & 0 & 0 & 1 \end{bmatrix}$$

$$A_2^1 = \begin{bmatrix} 0 & 0 & 1 & 0 \\ 1 & 0 & 0 & 0 \\ 0 & 1 & 0 & d2 \\ 0 & 0 & 0 & 1 \end{bmatrix} \quad \left(A_2^1\right)^{-1} = \begin{bmatrix} 0 & 1 & 0 & 0 \\ 0 & 0 & 1 & -d2 \\ 1 & 0 & 0 & 0 \\ 0 & 0 & 0 & 1 \end{bmatrix}$$

$$A_3^2 = \begin{bmatrix} 1 & 0 & 0 & 0 \\ 0 & 1 & 0 & 0 \\ 0 & 0 & 1 & d3 \\ 0 & 0 & 0 & 1 \end{bmatrix} \quad \left(A_3^2\right)^{-1} = \begin{bmatrix} 1 & 0 & 0 & 0 \\ 0 & 1 & 0 & 0 \\ 0 & 0 & 1 & -d3 \\ 0 & 0 & 0 & 1 \end{bmatrix}$$

$$A_4^3 = \begin{bmatrix} C4 & -S4 & 0 & 0 \\ S4 & C4 & 0 & 0 \\ 0 & 0 & 1 & La \\ 0 & 0 & 0 & 1 \end{bmatrix} \quad T = \begin{bmatrix} nx & sx & ax & px \\ ny & sy & ay & py \\ nz & sz & az & pz \\ 0 & 0 & 0 & 1 \end{bmatrix}$$

A careful approach leads to the development of both sides of the equation 6.4:

$$\begin{bmatrix} -S1nx + C1ny & -S1sx + C1sy & -S1ax + C1ay & -S1px + C1py \\ nz & sz & az & pz - d_2 \\ C1nx + S1ny & C1sx + S1sy & C1ax + S1ay & C1px + S1py - d_3 \\ 0 & 0 & 0 & 1 \end{bmatrix} = \begin{bmatrix} C4 & -S4 & 0 & 0 \\ S4 & C4 & 0 & 0 \\ 0 & 0 & 1 & La \\ 0 & 0 & 0 & 1 \end{bmatrix}$$

Comparing the first three elements of the last column, on both sides of the equality:

$$-S1px + C1py = 0 \qquad (6.5)$$

$$pz - d_2 = 0 \qquad (6.6)$$

$$C1px + S1py - d_3 = La \qquad (6.7)$$

A set of four equations (necessary to get four unknown variables: θ_1, d_2, d_3, θ_4) may be obtained matching the elements of the second row, first column:

$$nz = S4 \qquad (6.8)$$

From the equations 6.5 and 6.6 is easy reproduce the results of θ_1 and d_2.

To solve the equation 6.7, first it proceeds to get expressions of S1 and C1 in function of *px* and *py*. A rectangle triangle as the figure 6.6 is useful for this purpose.

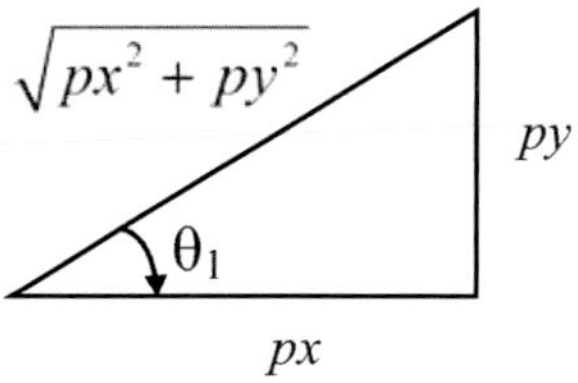

$$S1 = \frac{py}{\sqrt{px^2 + py^2}}$$

$$C1 = \frac{px}{\sqrt{px^2 + py^2}}$$

Figure 6.6 Trigonometric relationships of *px*, *py* with θ_1 in a Cylindrical Robot.

Now the expressions of S1 and C1 are included in the equation 6.7. After rationalize, it will be checked again the result obtained and shown to d_3.

Finally, isolating the angle from the equation 6.8 it will obtain the indicated for θ_4.

6.7. DIFFERENTIAL KINEMATICS AND SINGULARITIES

We proceed to determine the Jacobian matrix, its determinant and will conclude with the calculation of singularities, based on the equations of position.

$$px = C1(La + d_3)$$
$$py = S1(La + d_3)$$
$$pz = d_2$$

$$J = \begin{bmatrix} -Sin\theta_1(La + d_3) & 0 & Cos\theta_1 \\ Cos\theta_1(La + d_3) & 0 & Sin\theta_1 \\ 0 & 1 & 0 \end{bmatrix}$$

$$|J| = \left(Cos\theta_1\right)^2 (La + d_3) + \left(Sin\theta_1\right)^2 (La + d_3) = 0$$

The above equality only is possible if d_3=-La. If it were possible to reproduce this value for the parameter of the third joint it happened that *px=py=0*, which would complete the circular section of cylindrical Work Volume, that characterizes this robot.

CHAPTER 7

KINEMATICS OF SPHERICAL ROBOT

"It is not enough to take steps which may some day lead to a goal; each step must be itself a goal and a step likewise".

Johann Wolfgang Von Goethe - Johann P. Eckermann

7.1. FUNCTIONAL REPRESENTATION AND COORDINATE SYSTEMS

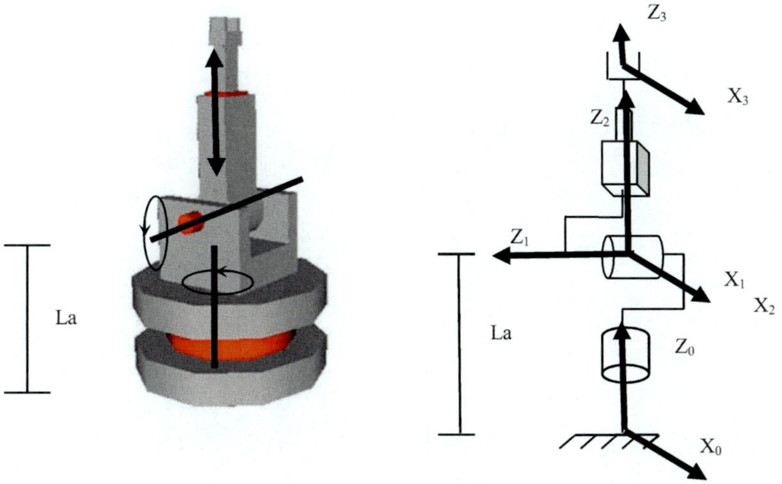

Figure 7.1 Functional representation and generalized coordinates systems of Spherical Robot

7.2. TABLE OF DH PARAMETERS

Table 7.1 DH parameters of Spherical Robot

Joint	θ	d	a	α	q
1	q_1	La	0	90	θ_1
2	q_2	0	0	-90	θ_2
3	0	q_3	0	0	d_3

7.3 TRANSITION HOMOGENEOUS MATRICES

General equation:

$$^{i-1}A_i = \begin{bmatrix} C\theta & -C\alpha S\theta & S\alpha S\theta & aC\theta \\ S\theta & C\alpha C\theta & -S\alpha C\theta & aS\theta \\ 0 & S\alpha & C\alpha & d \\ 0 & 0 & 0 & 1 \end{bmatrix}$$

Particular matrices:

$$A_1^0 = \begin{bmatrix} C1 & 0 & S1 & 0 \\ S1 & 0 & -C1 & 0 \\ 0 & 1 & 0 & La \\ 0 & 0 & 0 & 1 \end{bmatrix} \quad A_2^1 = \begin{bmatrix} C2 & 0 & -S2 & 0 \\ S2 & 0 & C2 & 0 \\ 0 & -1 & 0 & 0 \\ 0 & 0 & 0 & 1 \end{bmatrix}$$

$$A_3^2 = \begin{bmatrix} 1 & 0 & 0 & 0 \\ 0 & 1 & 0 & 0 \\ 0 & 0 & 1 & d3 \\ 0 & 0 & 0 & 1 \end{bmatrix}$$

7.4. TRANSFORMATION MATRIX OF THE ROBOT

$$T = A_1^0 * A_2^1 * A_3^2$$

$$T = \begin{bmatrix} C1C2 & -S1 & -C1S2 & -d3C1S2 \\ S1C2 & C1 & -S1S2 & -d3S1S2 \\ S2 & 0 & C2 & d3C2 + La \\ 0 & 0 & 0 & 1 \end{bmatrix}$$

7.5. GEOMETRIC STUDY

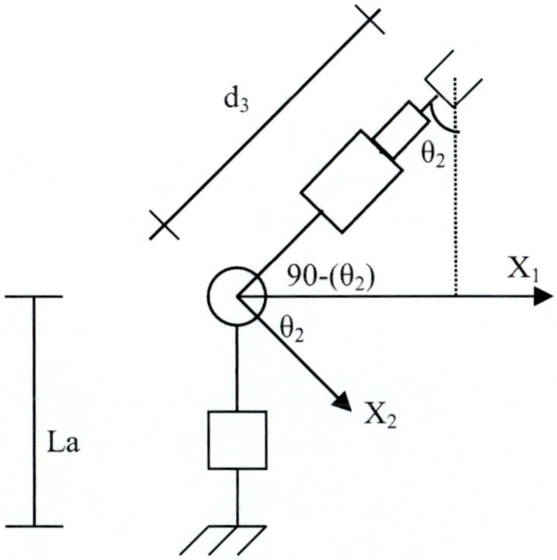

Figure 7.2 Side view of the Spherical Robot in the fourth quadrant.

In the side view of Figure 7.2, into the triangle is completed by the dotted line, the adjacent side having a length equal to $d_3\text{Cos}(90\text{-}\theta_2)$, which is equivalent to $d_3\text{Sin}(\theta_2)$.

Then, in an overhead view as shown in Figure 7.3, you can relate to θ_1 with the position components px, py.

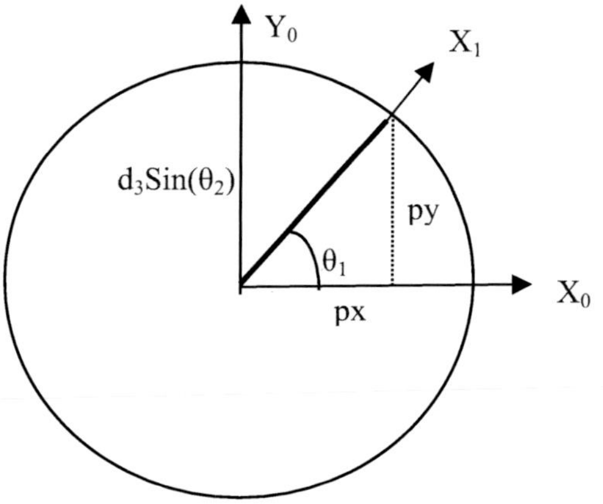

Figure 7.3 Overhead view of Spherical robot in the first quadrant

It is easy deducing the searched relations between the position (px, py) and the turn radius, $d_3Sin(\theta_2)$:

$$px = d_3 C1S2$$
$$py = d_3 S1S2$$

These expressions are equivalent to those obtained in the matrix T. They come from an analysis where θ_2 is negative (fourth quadrant); and if this sign is made explicit, you get to $Sin(-\theta_2) = - Sin(\theta_2)$.

Respect to Z component position of TCP, in the figure 7.2, from the fixed base of the robot, where the origin of the system $X_0Y_0Z_0$ is, it can be deduced that:

$$pz = La + d_3Cos(\theta_2)$$

This is the same expression into the matrix T, considering that $Cos(-\theta_2)$ is equal to $Cos(\theta_2)$.

7.6. ANALYSIS OF RESULTS

a) $\theta_2 = -90^o$, $\theta_1 = \theta_3 = 0^o$

The posed condition is shown in Figure 7.4.

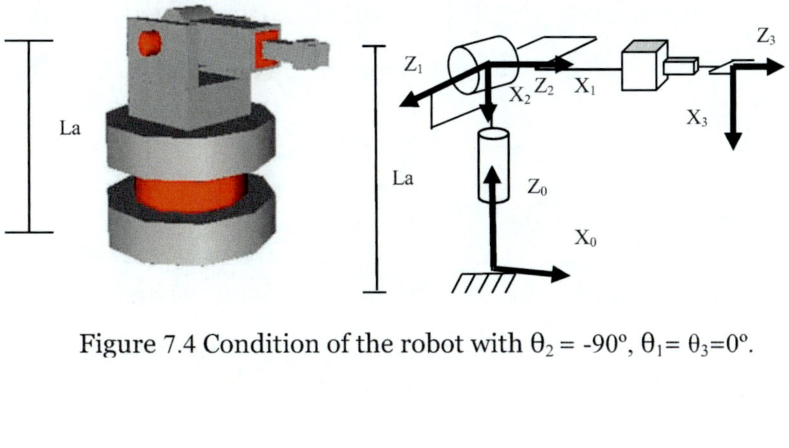

Figure 7.4 Condition of the robot with $\theta_2 = -90^o$, $\theta_1 = \theta_3 = 0^o$.

Easy is getting by inspection that the correct values of TCP position are: $px = d_3$, $py = 0$, $pz =$ La. This allows testing the model obtained above. Indeed, it evaluating the position equations, we have:

$$px = -d_3 C1S2 = d_3 Cos(0)Sin(-90) = d_3$$
$$py = -d_3 S1S2 = -d_3 Sin(0)Sin(-90) = 0$$
$$pz = La + d_3 Cos(-\theta_2) = La + d_3 Cos(90) = La$$

b) $\theta_1 = \theta_2 = \theta_3 = 0^o$

It proceeds to evaluate again the previous equations.

$$px = -d_3 C1S2 = d_3 Cos(0)Sin(0) = 0$$
$$py = -d_3 S1S2 = -d_3 Sin(0)Sin(0) = 0$$
$$pz = La + d_3 Cos(-\theta_2) = La + d_3 Cos(0) = La + d_3$$

The result is absolutely correct, in the light of what can be observed on the functional representation of the first case.

7.7. INVERSE KINEMATICS

Once again it is begins writing the matrix equations using the Transition Homogeneous Matrices:

$$T = A_1^0 * A_2^1 * A_3^2$$
$$\left(A_1^0\right)^{-1} * T = A_2^1 * A_3^2 \tag{7.1}$$
$$\left(A_2^1\right)^{-1} * \left(A_1^0\right)^{-1} * T = A_3^2$$

From the equation 7.1:

$$\begin{bmatrix} C1 & S1 & 0 & 0 \\ 0 & 0 & 1 & -La \\ S1 & -C1 & 0 & 0 \\ 0 & 0 & 0 & 1 \end{bmatrix} \begin{bmatrix} nx & sx & ax & px \\ ny & sy & ay & py \\ nz & sz & az & pz \\ 0 & 0 & 0 & 1 \end{bmatrix} =$$

$$\begin{bmatrix} C2 & 0 & -S2 & 0 \\ S2 & 0 & C2 & 0 \\ 0 & -1 & 0 & 0 \\ 0 & 0 & 0 & 1 \end{bmatrix} \begin{bmatrix} 1 & 0 & 0 & 0 \\ 0 & 1 & 0 & 0 \\ 0 & 0 & 1 & d3 \\ 0 & 0 & 0 & 1 \end{bmatrix}$$

$$\begin{bmatrix} nxC1 + nyS1 & sxC1 + syS1 & axC1 + ayS1 & pxC1 + pyS1 \\ nz & sz & az & pz - La \\ nxS1 - nyC1 & sxS1 - syC1 & axS1 - ayC1 & pxS1 - pyC1 \\ 0 & 0 & 0 & 1 \end{bmatrix} =$$

$$\begin{bmatrix} C2 & 0 & -S2 & d3S2 \\ S2 & 0 & C2 & d3C2 \\ 0 & -1 & 0 & 0 \\ 0 & 0 & 0 & 1 \end{bmatrix}$$

The goal is to determine θ_1, θ_2 and d_3 as functions of px, py and pz. To achieve this, we started comparing the elements of the third row, fourth column (3,4) on both sides of the equation:

$$pxS1 - pyC1 = 0$$

With the immediate result:

$$\boxed{\theta_1 = Tan^{-1}(py / px)}$$

Since θ_1 is known, is possible use it into the brief two equations system, which was obtained posing the equality between the entries (1,4) and (2,4):

$$pxC1 + pyS1 = d_3S2 \quad (7.2)$$

$$pz - La = d_3C2 \quad (7.3)$$

It is used here the resource of a triangle for get expressions of S1 and C1, to use them in Equation 7.2.

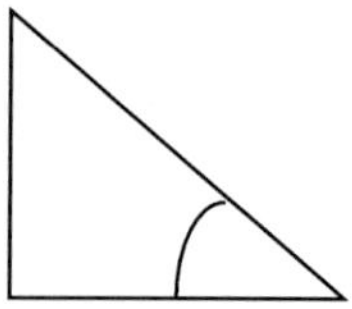

$$S1 = py / \sqrt{px^2 + py^2}$$

$$C1 = px / \sqrt{px^2 + py^2}$$

$$\frac{px^2 + py^2}{\sqrt{px^2 + py^2}} = d_3S2 \Rightarrow \sqrt{px^2 + py^2} = d_3S2$$

This variant of equation 7.2 is divided by the equation 7.3 to remove d3 and determine θ_2.

$$\frac{\sqrt{px^2+py^2}}{pz-La}=Tan(\theta_2) \Rightarrow \boxed{\theta_2=Tan^{-1}\left(\frac{\sqrt{px^2+py^2}}{pz-La}\right)}$$

Again the same resource of the triangle is used to obtain an expression of $Cos\theta_2$ and use it in equation 7.3.

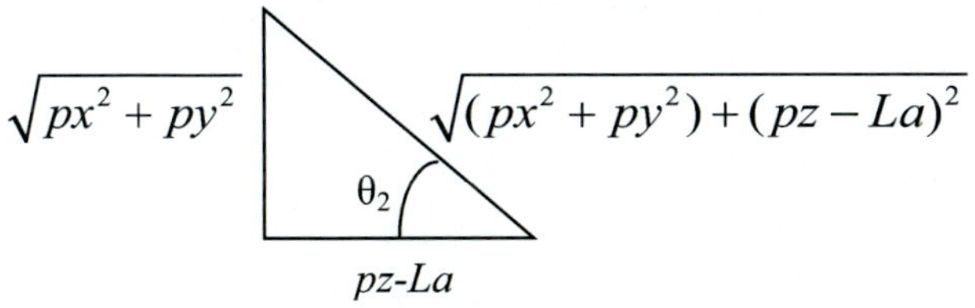

$$C2=\frac{pz-La}{\sqrt{(px^2+py^2)+(pz-La)}}$$

$$pz-La=d_3\frac{pz-La}{\sqrt{(px^2+py^2)+(pz-La)^2}} \Rightarrow$$

$$\boxed{d_3=\sqrt{(px^2+py^2)+(pz-La)^2}}$$

Equivalent solutions are obtained much more easily by analyzing the robot in a spherical coordinate system with origin in the second revolute joint (Figure 7.5).

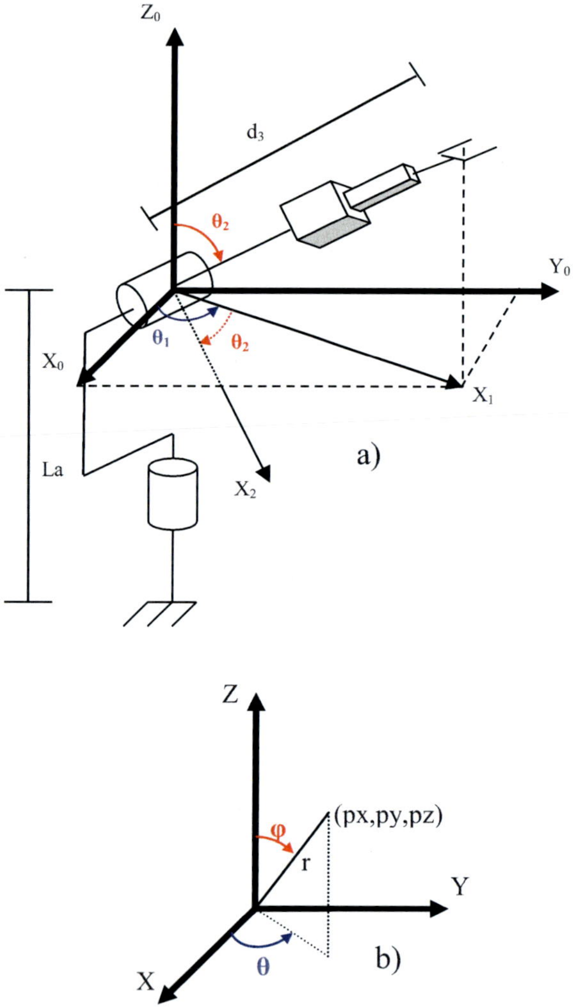

Figure 7.5 a) Representation of Spherical Robot adapted to analysis in spherical coordinates. b) System of spherical

The expression in coordinate Cartesian system (px, py, pz) has an equivalent in spherical coordinates (r, θ, φ) for any point, which it can be got using the following equations

$$r = \sqrt{px^2 + py^2 + pz^2} \tag{7.4}$$

$$\theta = Tan^{-1}(py / px) \tag{7.5}$$

$$\phi = Sin^{-1}\left(\frac{\sqrt{px^2 + py^2}}{\sqrt{px^2 + py^2 + pz^2}} \right) \tag{7.6}$$

For the present case it is only relevant to establish the following correspondences:

$$\boxed{r = d_3 = \sqrt{px^2 + py^2 + pz^2}}$$

$$\boxed{\theta = \theta_1 = Tan^{-1}(py / px)}$$

$$\boxed{\phi = \theta_2 = Sin^{-1}\left(\frac{\sqrt{px^2 + py^2}}{\sqrt{px^2 + py^2 + pz^2}} \right)}$$

These results are completely correct, understanding that there has been a displacement of the origin of the system $X_0Y_0Z_0$ in "*La*" units of length, so that *pz* is equal to the previous *pz* minus "*La*".

Respect to θ_2 there is no difference because the expression in function of the inverse Sinus is equivalent to the previous which was made in function of the inverse Tangent.

CHAPTER 8

KINEMATICS OF SCARA ROBOT

"Said the dog to the bone: If you're hard, I have time"

Anonymous

The SCARA Robot studied in this chapter is like to the shown in the figure 8.1. It was developed as the product of two degree thesis, between the Schools of Mechanical and Electrical Engineering of University of Carabobo (Valencia, Venezuela). It was initially conceived as a tool for marking metal sheets, with the Workspace shown in Figure 8.2. It is currently at the Laboratory of Robotics and Industrial Vision of the Faculty of Engineering of that University.

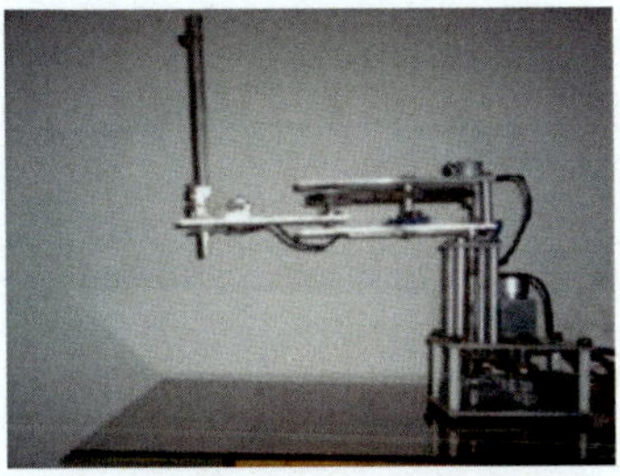

Figure 8.1 SCARA Robot University of Carabobo
From: Rivas, Adolfo, 2005

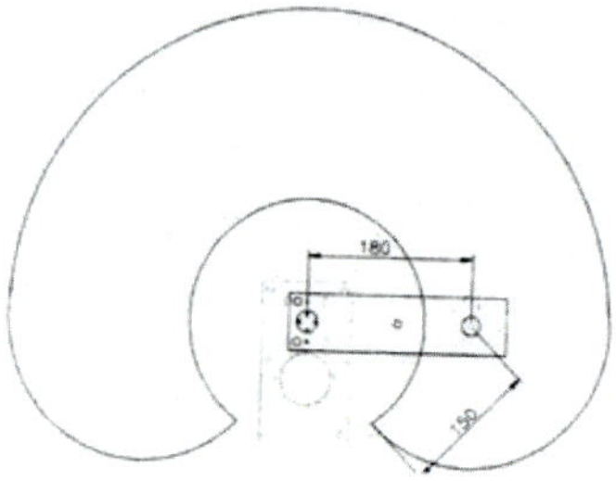

Figure 8.2 Workspace of SCARA Robot, University of Carabobo
From: Froment – Sánchez, 2004

8.1. FUNCTIONAL REPRESENTATION AND COORDINATE SYSTEMS

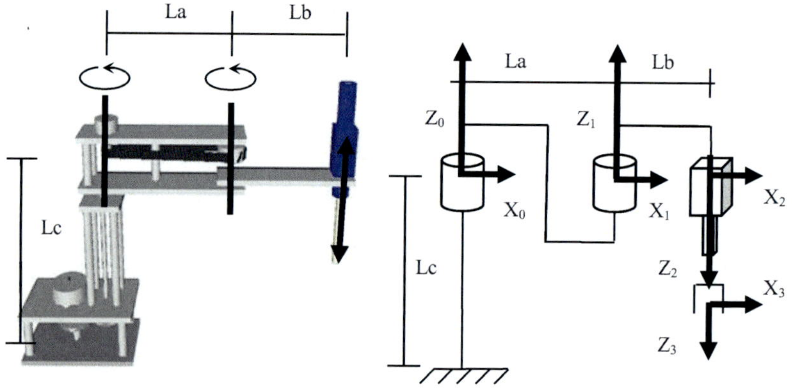

Figure 8.3 Functional representation and generalized coordinates systems of SCARA Robot.
From: Rivas, Adolfo (2005)

8.2. TABLE OF DH PARAMETERS

Table 8.1 DH Parameters of SCARA Robot

Joint	θ	d	a	α	q
1	q_1	0	La	0	θ_1
2	q_2	0	Lb	180	θ_2
3	0	q_3	0	0	d_3

8.3. TRANSITION HOMOGENEOUS MATRICES

General equation:

$$
{}^{i-1}A_i = \begin{bmatrix} C\theta & -C\alpha S\theta & S\alpha S\theta & aC\theta \\ S\theta & C\alpha C\theta & -S\alpha C\theta & aS\theta \\ 0 & S\alpha & C\alpha & d \\ 0 & 0 & 0 & 1 \end{bmatrix}
$$

Particular matrices:

$$
A_1^0 = \begin{bmatrix} C1 & -S1 & 0 & LaC1 \\ S1 & C1 & 0 & LaS1 \\ 0 & 0 & 1 & 0 \\ 0 & 0 & 0 & 1 \end{bmatrix} \quad A_2^1 = \begin{bmatrix} C2 & S2 & 0 & LbC2 \\ S2 & -C2 & 0 & LbS2 \\ 0 & 0 & -1 & 0 \\ 0 & 0 & 0 & 1 \end{bmatrix}
$$

$$
A_3^2 = \begin{bmatrix} 1 & 0 & 0 & 0 \\ 0 & 1 & 0 & 0 \\ 0 & 0 & 1 & d3 \\ 0 & 0 & 0 & 1 \end{bmatrix}
$$

8.4. TRANSFORMATION MATRIX OF THE ROBOT

$$T = A_1^0 * A_2^1 * A_3^2$$

$$T = \begin{bmatrix} C1C2 - S1S2 & C1S2 + S1C2 & 0 & Lb\left(C1C2 - S1S2\right) + LaC1 \\ S1C2 + C1S2 & S1S2 - C1C2 & 0 & Lb\left(S1C2 + C1S2\right) + LaS1 \\ 0 & 0 & -1 & -d3 \\ 0 & 0 & 0 & 1 \end{bmatrix}$$

8.5. ALTERNATIVE STUDIES

8.5.1. Reference at bottom

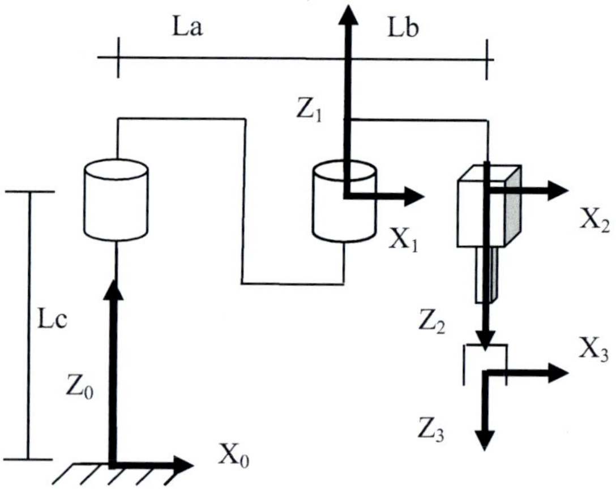

Figure 8.4 Functional representation and generalized coordinates systems of SCARA Robot, with reference at the Base

Table 8.2 DH Parameters of SCARA Robot with reference at the Base

Joint	θ	d	a	α	q
1	q_1	Lc	La	0	θ_1
2	q_2	0	Lb	180	θ_2
3	0	q_3	0	0	d_3

$$A_1^0 = \begin{bmatrix} C1 & -S1 & 0 & LaC1 \\ S1 & C1 & 0 & LaS1 \\ 0 & 0 & 1 & Lc \\ 0 & 0 & 0 & 1 \end{bmatrix} \quad A_2^1 = \begin{bmatrix} C2 & S2 & 0 & LbC2 \\ S2 & -C2 & 0 & LbS2 \\ 0 & 0 & -1 & 0 \\ 0 & 0 & 0 & 1 \end{bmatrix}$$

$$A_3^2 = \begin{bmatrix} 1 & 0 & 0 & 0 \\ 0 & 1 & 0 & 0 \\ 0 & 0 & 1 & d3 \\ 0 & 0 & 0 & 1 \end{bmatrix}$$

$$T = \begin{bmatrix} C1C2 - S1S2 & C1S2 + S1C2 & 0 & Lb(C1C2 - S1S2) + LaC1 \\ S1C2 + C1S2 & S1S2 - C1C2 & 0 & Lb(S1C2 + C1S2) + LaS1 \\ 0 & 0 & -1 & Lc - d3 \\ 0 & 0 & 0 & 1 \end{bmatrix}$$

It can be seen that the only difference from the previous case is the Z component of the TCP position: pz = Lc-d3, including additional constant Lc as a logical consequence of the displacement of the reference system S0.

8.5.2. Reference at bottom and Z_2, Z_3 pointing up

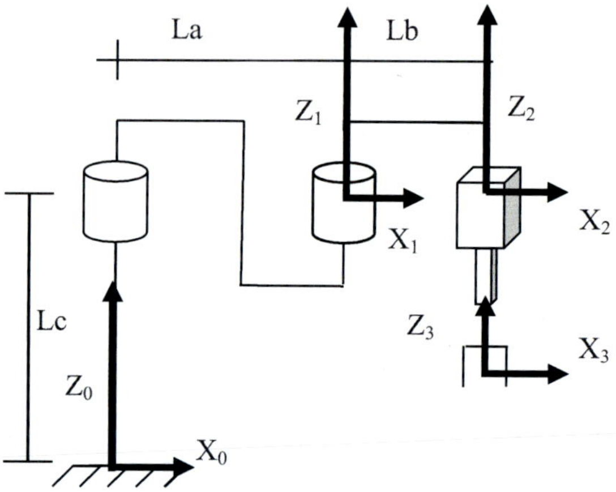

Figure 8.5 Functional representation and generalized coordinates systems of SCARA Robot, with reference at the Base and Z_2, Z_3 pointing up.

Table 8.3 DH Parameters of SCARA Robot with negative d_3

Joint	θ	d	a	α	q
1	q_1	Lc	La	0	θ_1
2	q_2	0	Lb	0	θ_2
3	0	q_3	0	0	$-d_3$

This variation in the selection of the directions of Z_2 and Z_3 axes produces the disappearance 180 degrees of the parameter α_2 and the necessary indication of minus sign at d_3, because the offsets provided for the prism are negatives. Such variations are evident in the changes of sign in the entries (1,2), (2,2) and (3,3) of the matrix $A^1{}_2$; and element (3,4) of the matrix $A^2{}_3$, corresponding to the negative parameter of the third.

$$A_1^0 = \begin{bmatrix} C1 & -S1 & 0 & LaC1 \\ S1 & C1 & 0 & LaS1 \\ 0 & 0 & 1 & Lc \\ 0 & 0 & 0 & 1 \end{bmatrix} \quad A_2^1 = \begin{bmatrix} C2 & -S2 & 0 & LbC2 \\ S2 & C2 & 0 & LbS2 \\ 0 & 0 & 1 & 0 \\ 0 & 0 & 0 & 1 \end{bmatrix}$$

$$A_3^2 = \begin{bmatrix} 1 & 0 & 0 & 0 \\ 0 & 1 & 0 & 0 \\ 0 & 0 & 1 & -d3 \\ 0 & 0 & 0 & 1 \end{bmatrix}$$

$$T = \begin{bmatrix} C1C2 - S1S2 & -C1S2 - S1C2 & 0 & Lb\left(C1C2 - S1S2\right) + LaC1 \\ S1C2 + C1S2 & -S1S2 + C1C2 & 0 & Lb\left(S1C2 + C1S2\right) + LaS1 \\ 0 & 0 & 1 & Lc - d3 \\ 0 & 0 & 0 & 1 \end{bmatrix}$$

It is therefore evident that the changes made for this study do not alter the equations of TCP position.

8.6. GEOMETRIC STUDY

This analysis is done from the perspective of a side view (figure 8.6) and an overhead view (figures 8.7 and 8.8).

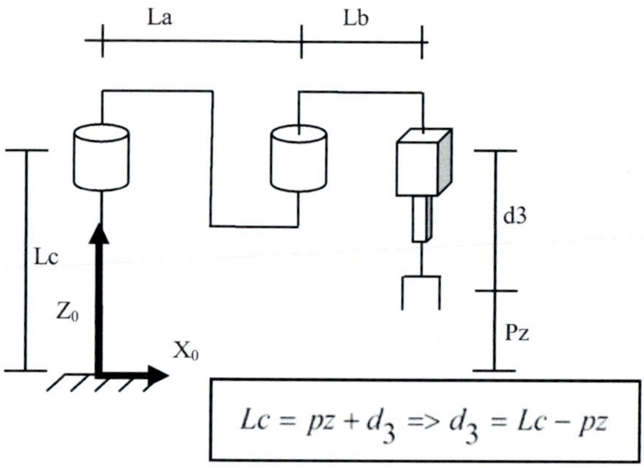

Figure 8.6 Side view of SCARA robot. Getting of d_3 parameter.

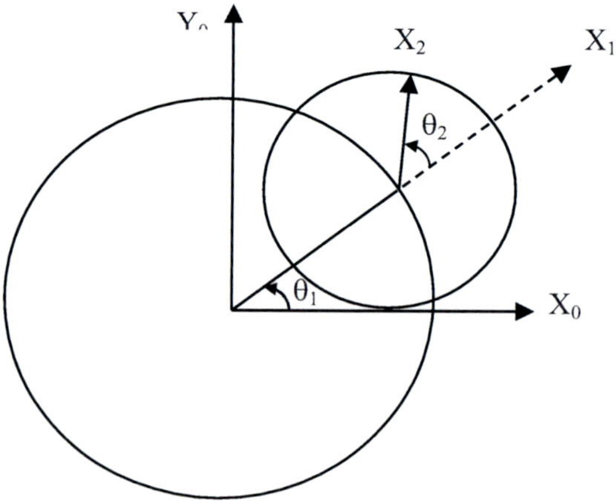

Figure 8.7 Overhead view of SCARA robot.

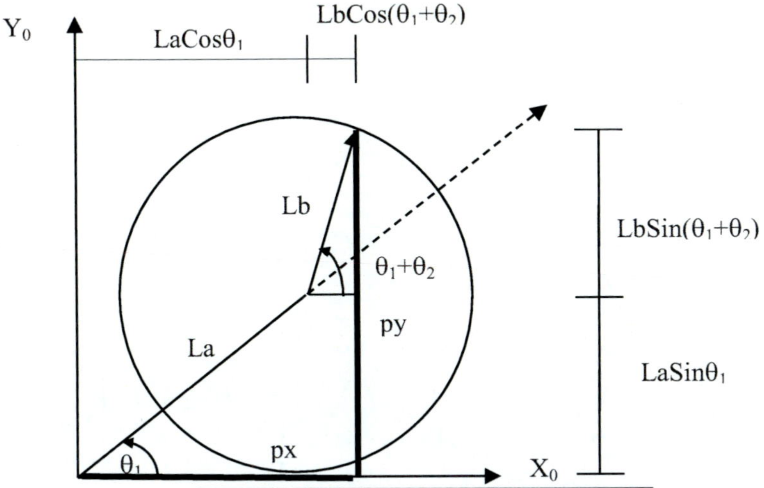

$$px = LaCos\theta_1 + LbCos(\theta_1 + \theta_2)$$

$$py = LaSin\theta_1 + LbSin(\theta_1 + \theta_2)$$

Figure 8.8 Geometric analysis of SCARA robot to getting of parameters θ_1 y θ_2.

8.7. INVERSE KINEMATICS

It is start with the Transition Homogeneous matrices obtained in Section 8.3 of this chapter, to pose the following equations:

$$\begin{aligned} T &= A_1^0 * A_2^1 * A_3^2 \\ \left(A_1^0\right)^{-1} * T &= A_2^1 * A_3^2 \\ \left(A_2^1\right)^{-1} * \left(A_1^0\right)^{-1} * T &= A_3^2 \end{aligned} \qquad (8.1)$$

Developing the equation 8.1:

$$\begin{bmatrix} C1 & S1 & 0 & -La \\ -S1 & C1 & 0 & 0 \\ 0 & 0 & 1 & 0 \\ 0 & 0 & 0 & 1 \end{bmatrix} \begin{bmatrix} nx & sx & ax & px \\ ny & sy & ay & py \\ nz & sz & az & pz \\ 0 & 0 & 0 & 1 \end{bmatrix} =$$

$$\begin{bmatrix} C2 & S2 & 0 & LbC2 \\ S2 & -C2 & 0 & LbS2 \\ 0 & 0 & -1 & 0 \\ 0 & 0 & 0 & 1 \end{bmatrix} \begin{bmatrix} 1 & 0 & 0 & 0 \\ 0 & 1 & 0 & 0 \\ 0 & 0 & 1 & d3 \\ 0 & 0 & 0 & 1 \end{bmatrix}$$

$$\begin{bmatrix} nxC1 + nyS1 & sxC1 + syS1 & axC1 + ayS1 & pxC1 + pyS1 - La \\ -nxS1 + nyC1 & -sxS1 + syC1 & -axS1 + ayC1 & -pxS1 + pyC1 \\ nz & sz & az & pz \\ 0 & 0 & 0 & 1 \end{bmatrix}$$

$$\begin{bmatrix} C2 & S2 & 0 & LbC2 \\ S2 & -C2 & 0 & LbS2 \\ 0 & 0 & -1 & -d3 \\ 0 & 0 & 0 & 1 \end{bmatrix}$$

From comparison of the elements of the last column in each array in both sides of the equation, the following equations are obtained:

$$C1px + S1py - La = LbC2 \qquad (8.2)$$

$$-S1px + C1py = LbS2 \qquad (8.3)$$

$$d3 = -pz$$

A possible way of solution is the following:

$$C1px + S1py - La = LbC2$$
$$-S1px + C1py \qquad = LbS2$$

Doing square to both sides of the equations and adding them:

$$(C1px + S1py)^2 - 2La(C1px + S1py) + La^2 = Lb^2C2^2$$
$$(-S1px + C1py)^2 \qquad = Lb^2S2^2$$
$$\overline{(px^2 + py^2) - 2La(C1px + S1py) + La^2 = Lb^2}$$

Simplifying:

$$\frac{(px^2 + py^2) + La^2 - Lb^2}{2La} = (C1px + S1py)$$

Let us doing the following definition:

$$Sea: \; K = \frac{(px^2 + py^2) + La^2 - Lb^2}{2La}$$
$$C1px + S1py = K$$

Now the Sinθ is replaced using a trigonometric identity:

$$\pm\left(\sqrt{1-C1^2}\right)py = K - C1px$$

Again, the square of both sides followed of a reordering for posing a quadratic equation where C1 is the variable:

$$(1-C1^2)py^2 = K^2 - 2KpxC1 + px^2C1^2$$

$$(px^2+py^2)C1^2 - 2KpxC1 + (K^2-py^2) = 0$$

Finally, it is possible isolating C1and then, θ_1 and θ_2.

$$C1 = \frac{2Kpx \pm \sqrt{4K^2px^2 - 4(px^2+py^2)(K^2-py^2)}}{2(px^2+py^2)}$$

$$C1 = \frac{Kpx \pm py\sqrt{px^2+py^2-K^2}}{(px^2+py^2)}$$

$$\pm\theta_1 = \begin{bmatrix} \cos^{-1}\left(\dfrac{Kpx + py\sqrt{px^2+py^2-K^2}}{(px^2+py^2)}\right) \\ \cos^{-1}\left(\dfrac{Kpx - py\sqrt{px^2+py^2-K^2}}{(px^2+py^2)}\right) \end{bmatrix}$$

$$\theta_2 = \begin{bmatrix} \sin^{-1}\left(\dfrac{-S1px + C1py}{Lb}\right) \\ \pi - \sin^{-1}\left(\dfrac{-S1px + C1py}{Lb}\right) \end{bmatrix}$$

These equations allow calculate the values of both angles to reach any position defined by *px* and *py*. However, just is noting that there are eight (8) possible combinations of θ_1 and θ_2; and only two of them are physically realizable.

An alternative way consists on appealing to specialized software to solve the system of equations (8.2) and (8.3). The following is a proven solution:

```
th1=[atan2(-
1/2*(px/(4*py^2+4*px^2)*(4*px*La^2+4*px*py^2-
4*px*Lb^2+4*px^3+4*(-py^2*Lb^4-2*px^2*py^4-px^4*py^2-
py^2*La^4+2*py^4*La^2+2*py^4*Lb^2-
py^6+2*px^2*La^2*py^2+2*px^2*py^2*Lb^2+2*py^2*La^2*Lb^2)^(1/2))-
py^2-px^2-
La^2+Lb^2)/La/py,1/2/(4*py^2+4*px^2)*(4*px*La^2+4*px*py^2-
4*px*Lb^2+4*px^3+4*(-py^2*Lb^4-2*px^2*py^4-px^4*py^2-
py^2*La^4+2*py^4*La^2+2*py^4*Lb^2-
py^6+2*px^2*La^2*py^2+2*px^2*py^2*Lb^2+2*py^2*La^2*Lb^2)^(1/2))/La)
*180/pi;

atan2(-1/2*(px/(4*py^2+4*px^2)*(4*px*La^2+4*px*py^2-
4*px*Lb^2+4*px^3-4*(-py^2*Lb^4-2*px^2*py^4-px^4*py^2-
py^2*La^4+2*py^4*La^2+2*py^4*Lb^2-
py^6+2*px^2*La^2*py^2+2*px^2*py^2*Lb^2+2*py^2*La^2*Lb^2)^(1/2))-
```

```
py^2-px^2-
La^2+Lb^2)/La/py,1/2/(4*py^2+4*px^2)*(4*px*La^2+4*px*py^2-
4*px*Lb^2+4*px^3-4*(-py^2*Lb^4-2*px^2*py^4-px^4*py^2-
py^2*La^4+2*py^4*La^2+2*py^4*Lb^2-
py^6+2*px^2*La^2*py^2+2*px^2*py^2*Lb^2+2*py^2*La^2*Lb^2)^(1/2))/La)
*180/pi];
th2=[atan2(1/2*(px^2/(4*py^2+4*px^2)*(4*px*La^2+4*px*py^2
-4*px*Lb^2+4*px^3+4*(-py^2*Lb^4-2*px^2*py^4-px^4*py^2-
py^2*La^4+2*py^4*La^2+2*py^4*Lb^2-
py^6+2*px^2*La^2*py^2+2*px^2*py^2*Lb^2+2*py^2*La^2*Lb^2)^(1/2))-
px*py^2-px^3-
px*La^2+px*Lb^2+py^2/(4*py^2+4*px^2)*(4*px*La^2+4*px*py^2-
4*px*Lb^2+4*px^3+4*(-py^2*Lb^4-2*px^2*py^4-px^4*py^2-
py^2*La^4+2*py^4*La^2+2*py^4*Lb^2-
py^6+2*px^2*La^2*py^2+2*px^2*py^2*Lb^2+2*py^2*La^2*Lb^2)^(1/2)))/La
/py/Lb,-1/2*(-py^2-px^2+La^2+Lb^2)/La/Lb)*180/pi;

atan2(1/2*(px^2/(4*py^2+4*px^2)*(4*px*La^2+4*px*py^2-
4*px*Lb^2+4*px^3-4*(-py^2*Lb^4-2*px^2*py^4-px^4*py^2-
py^2*La^4+2*py^4*La^2+2*py^4*Lb^2-
py^6+2*px^2*La^2*py^2+2*px^2*py^2*Lb^2+2*py^2*La^2*Lb^2)^(1/2))-
px*py^2-px^3-
px*La^2+px*Lb^2+py^2/(4*py^2+4*px^2)*(4*px*La^2+4*px*py^2-
4*px*Lb^2+4*px^3-4*(-py^2*Lb^4-2*px^2*py^4-px^4*py^2-
py^2*La^4+2*py^4*La^2+2*py^4*Lb^2-
py^6+2*px^2*La^2*py^2+2*px^2*py^2*Lb^2+2*py^2*La^2*Lb^2)^(1/2)))/La
/py/Lb,-1/2*(-py^2-px^2+La^2+Lb^2)/La/Lb)*180/pi];
```

In these expressions, *th1* and *th2* corresponding to θ_1 and θ_2, respectively; while "*atan2*" is the way the software used specifies the calculation of the fourth quadrant Inverse Tangent.

It may also be noted that the above expressions correspond to two-element vectors for both angles. This can be better understood with the figure 8.9, where it is evident that it is possible to reach the same point in the X, Y plane by two combinations of angles θ_1 and θ_2.

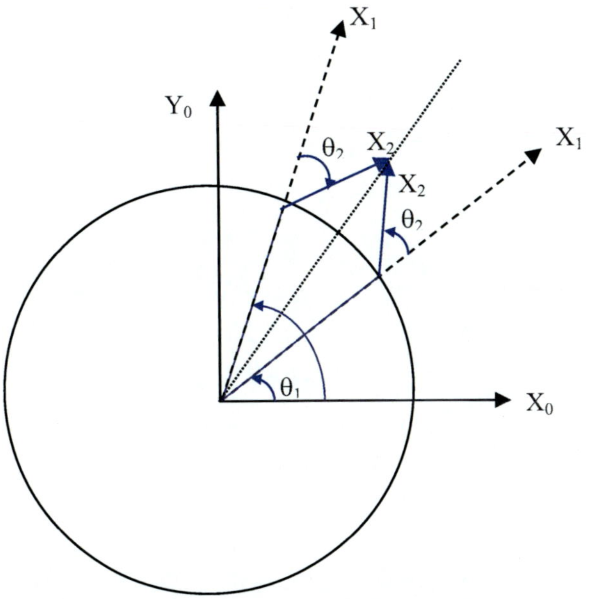

Figure 8.9 Two possible solutions for Inverse Kinematics of SCARA Robot

8.8. DIFFERENTIAL KINEMATICS

The position equations derived from the forward kinematics analysis allow getting a differential model base on Jacobian Matrix.

$$\begin{cases} px = LaCos\theta_1 + LbCos(\theta_1 + \theta_2) \\ py = LaSin\theta_1 + LbSin(\theta_1 + \theta_2) \\ pz = Lc - d_3 \end{cases}$$

$$\begin{cases} px' = \left[-LaSin\theta_1 - LbSin(\theta_1 + \theta_2)\right]\theta_1' - \left[LbSin\theta_2\right]\theta_2' + 0d_3' \\ py' = \left[LaCos\theta_1 + LbCos(\theta_1 + \theta_2)\right]\theta_1' + \left[LbCos\theta_2\right]\theta_2' + 0d_3' \\ pz' = 0\theta_1' + 0\theta_2' - d_3' \end{cases}$$

$$\boxed{\begin{bmatrix} px' \\ py' \\ pz' \end{bmatrix} = \begin{bmatrix} -LaSin\theta_1 - LbSin(\theta_1 + \theta_2) & -LbSin\theta_1 & 0 \\ LaCos\theta_1 + LbCos(\theta_1 + \theta_2) & LbCos\theta_1 & 0 \\ 0 & 0 & -1 \end{bmatrix} \begin{bmatrix} \theta_1' \\ \theta_2' \\ d_3' \end{bmatrix}}$$

The above result is the Direct Differential Kinematic Model of the SCARA robot; where de Jacobian Matrix may be used for get the values of joint parameters that determine the limits of Workspace of the robot. This is possible doing the calculus of the determinant of the Jacobian Matrix and equating it to zero.

$$|J| = \begin{vmatrix} -LaSin\theta_1 - LbSin(\theta_1 + \theta_2) & -LbSin\theta_1 & 0 \\ LaCos\theta_1 + LbCos(\theta_1 + \theta_2) & LbCos\theta_1 & 0 \\ 0 & 0 & -1 \end{vmatrix} = 0$$

$$LbCos\theta_1\left[LaSin\theta_1 + LbSin(\theta_1+\theta_2)\right] -$$
$$LbSin\theta_1\left[LaCos\theta_1 + LbCos(\theta_1+\theta_2)\right] = 0$$

$$LbCos\theta_1 LbSin(\theta_1+\theta_2) - LbSin\theta_1 LbCos(\theta_1+\theta_2) = 0$$

$$Cos\theta_1(Sin\theta_1Cos\theta_2 + Cos\theta_1Sin\theta_2) -$$
$$Sin\theta_1(Cos\theta_1Cos\theta_2 - Sin\theta_1Sin\theta_2) = 0$$

$$Cos\theta_1^2 Sin\theta_2 + Sin\theta_1^2 Sin\theta_2 = 0$$

$$(Cos\theta_1^2 + Sin\theta_1^2)Sin\theta_2 = 0$$

The above expression only is possible if θ_2 is null or 180º. With no restrictions for d_3 or θ_1, these values of θ_2 determine the inner and outer circles of the figure 8.10, where is defined the Workspace of robot in the XY plane.

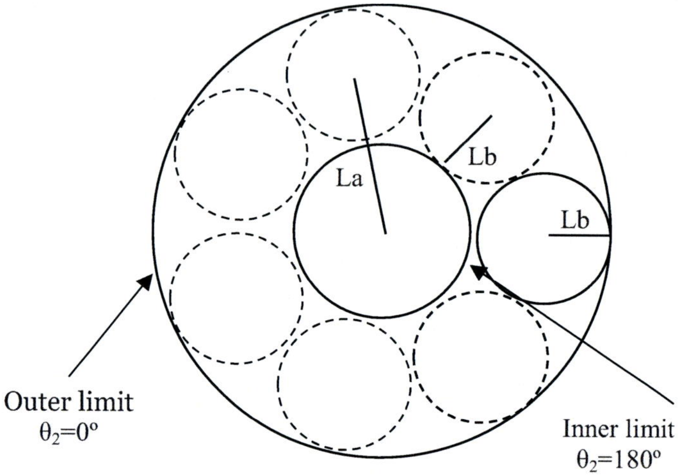

Figure 8.10 Workspace of SCARA Robot in the XY plane

CHAPTER 9

KINEMATICS OF ANTHROPOMORPHIC 5R ROBOT

"And the LORD God formed man of the dust of the ground, and breathed into his nostrils the breath of life; and man became a living soul".

Genesis, 2:7

"It would appear that we have reached the limits of what it's possible to achieve with computer technology, although one should be careful with such statements, as they tend to sound pretty silly in five years".

John Von Neumann (1949)

9.1. FUNCTIONAL REPRESENTATION AND COORDINATE SYSTEMS

Many industrial manipulators share this configuration (5R). The virtual model studied here was inspired by the Robot Initium built by Professor Jesús Pérez as part of his doctoral thesis in Instrumentation at the Faculty of Sciences of the Central University of Venezuela, under the guidance of Dr. Clemente Herrera.

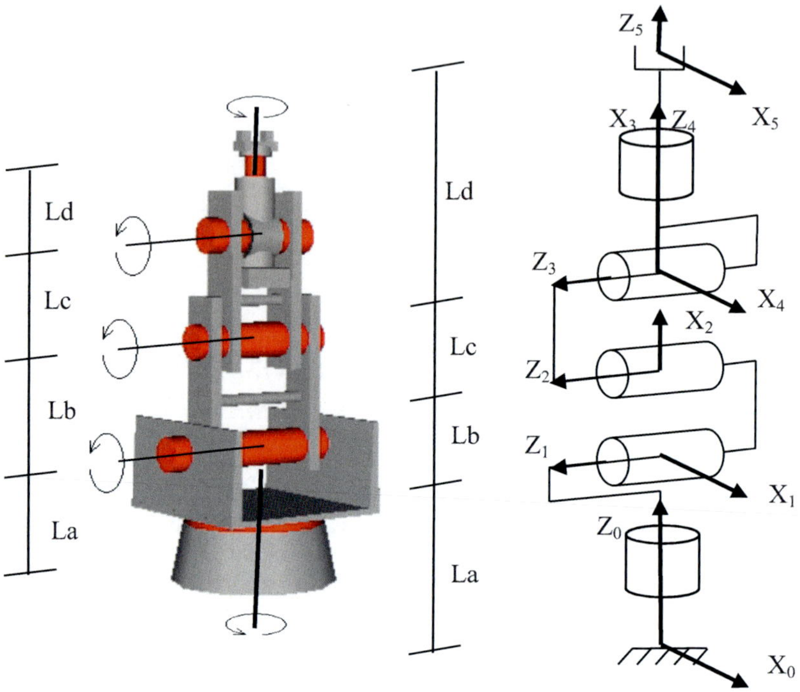

Figure 9.1 Functional representation and generalized coordinates systems of a 5R Robot

9.2. TABLE OF DH PARAMETERS

Table 9.1 DH Parameters of Initium Robot

Joint	θ	d	a	α	q
1	q_1	La	0	90	θ_1
2	q_2	0	Lb	0	θ_2
3	q_3	0	Lc	0	θ_3
4	q_4	0	0	-90	θ_4
5	q_5	Ld	0	0	θ_5

9.3. TRANSITION HOMOGENEOUS MATRICES

General equation:

$$^{i-1}A_i = \begin{bmatrix} C\theta & -C\alpha S\theta & S\alpha S\theta & aC\theta \\ S\theta & C\alpha C\theta & -S\alpha C\theta & aS\theta \\ 0 & S\alpha & C\alpha & d \\ 0 & 0 & 0 & 1 \end{bmatrix}$$

Particular matrices:

$$A_1^0 = \begin{bmatrix} C1 & 0 & S1 & 0 \\ S1 & 0 & -C1 & 0 \\ 0 & 1 & 0 & La \\ 0 & 0 & 0 & 1 \end{bmatrix} \quad A_2^1 = \begin{bmatrix} C2 & -S2 & 0 & LbC2 \\ S2 & C2 & 0 & LbS2 \\ 0 & 0 & 1 & 0 \\ 0 & 0 & 0 & 1 \end{bmatrix}$$

$$A_3^2 = \begin{bmatrix} C3 & -S3 & 0 & LcC3 \\ S3 & C3 & 0 & LcS3 \\ 0 & 0 & 1 & 0 \\ 0 & 0 & 0 & 1 \end{bmatrix} \quad A_4^3 = \begin{bmatrix} C4 & 0 & -S4 & 0 \\ S4 & 0 & C4 & 0 \\ 0 & -1 & 0 & 0 \\ 0 & 0 & 0 & 1 \end{bmatrix}$$

$$A_5^4 = \begin{bmatrix} C5 & -S5 & 0 & 0 \\ S5 & C5 & 0 & 0 \\ 0 & 0 & 1 & Ld \\ 0 & 0 & 0 & 1 \end{bmatrix}$$

9.4. ELEMENTS OF TRANSFORMATION MATRIX OF THE ROBOT

From the product of Transition Homogeneous matrices, the matrix T is obtained. Because of its size, are shown below its elements.

$$nx = C5(C4(C1C2C3 - C1S2S3) - S4(C1C2S3 + C1C3S2)) - S1S5$$
$$ny = C5(C4(C2C3S1 - S1S2S3) - S4(C2S1S3 + C3S1S2)) + C1S5$$
$$nz = C5(C4(C2S3 + C3S2) - S4(S2S3 - C2C3))$$

$$sx = - S5(C4(C1C2C3 - C1S2S3) - S4(C1C2S3 + C1C3S2)) - C5S1$$
$$sy = C1C5 - S5(C4(C2C3S1 - S1S2S3) - S4(C2S1S3 + C3S1S2))$$
$$sz = -S5(C4(C2S3 + C3S2) - S4(S2S3 - C2C3))$$

$$ax = - C4(C1C2S3 + C1C3S2) - S4(C1C2C3 - C1S2S3)$$
$$ay = - C4(C2S1S3 + C3S1S2) - S4(C2C3S1 - S1S2S3)$$
$$az = - C4(S2S3 - C2C3) - S4(C2S3 + C3S2)$$

$$px = C1C2Lb - Ld(C4(C1C2S3 + C1C3S2) + S4(C1C2C3 - C1S2S3)) + C1C2C3Lc - C1LcS2S3$$

$$py = C2LbS1 - Ld(C4(C2S1S3 + C3S1S2) + S4(C2C3S1 - S1S2S3)) + C2C3LcS1 - LcS1S2S3$$

$$pz = La - Ld(C4(S2S3 - C2C3) + S4(C2S3 + C3S2)) + LbS2 + C2LcS3 + C3LcS2$$

It is useful to apply appropriate trigonometric identities to the sum and difference of angles to simplify previous twelve expressions. In this process the direct kinematic model presented below is obtained.

$$nx = C1C(\theta_2+\theta_3+\theta_4)C5 - S1S5 \tag{9.1}$$
$$ny = S1C(\theta_2+\theta_3+\theta_4)C5 + C1S5 \tag{9.2}$$
$$nz = S(\theta_2+\theta_3+\theta_4)C5 \tag{9.3}$$

$$sx = -C1C(\theta_2+\theta_3+\theta_4)S5 - S1C5 \tag{9.4}$$
$$sy = -S1C(\theta_2+\theta_3+\theta_4)S5 + C1C5 \tag{9.5}$$
$$sz = -C(\theta_2+\theta_3+\theta_4)S5 \tag{9.6}$$

$$ax = -C1S(\theta_2+\theta_3+\theta_4) \tag{9.7}$$
$$ay = -S1S(\theta_2+\theta_3+\theta_4) \tag{9.8}$$
$$az = C(\theta_2+\theta_3+\theta_4) \tag{9.9}$$

$$px = C1\left[LbC2 + LcC(\theta_2+\theta_3) - LdS(\theta_2+\theta_3+\theta_4)\right] \tag{9.10}$$
$$py = S1\left[LbC2 + LcC(\theta_2+\theta_3) - LdS(\theta_2+\theta_3+\theta_4)\right] \tag{9.11}$$
$$pz = LdC(\theta_2+\theta_3+\theta_4) + LcS(\theta_2+\theta_3) + LbS2 + La \tag{9.12}$$

9.5. ANALYSIS OF RESULTS

The functional representation which is used to obtain the model (matrix T) shows a condition in which the robot $px = 0$, $py = 0$, $pz = La + Lb + Lc + Ld$. This position involves, as can be observed that the joints parameters are null ($\theta_1=\theta_3 = \theta_5 = 0^o$); except θ_2 and θ_4, whose values are 90° and -90°, respectively. Certainly, the position obtained by evaluating these values in the above equations coincides with expectations.

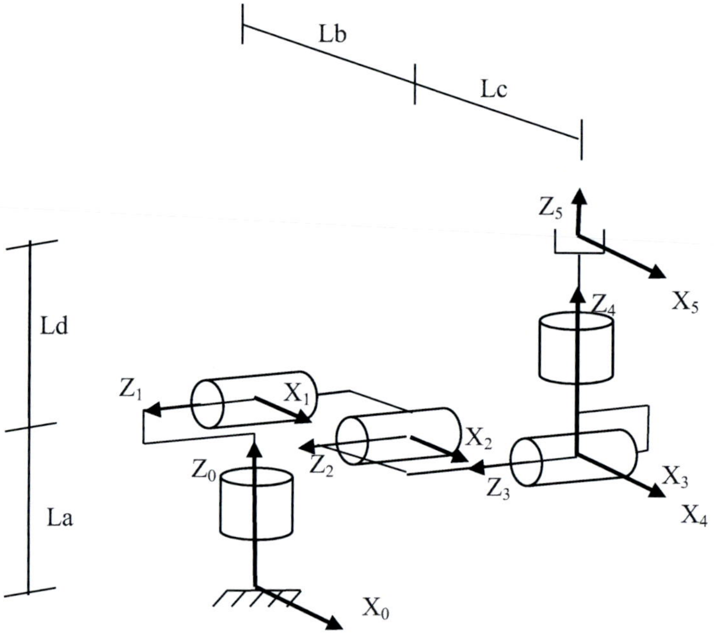

Figure 9.2 Condition for Initium Robot with all $\theta_i=0^o$.

When all joints parameters are nulls, the situation of the robot is as you can see in the figure 9.2.

When evaluating the position equations result: *px* = *Lb* + *Lc*, *py* = *0* and *pz* = *La* + *Ld*; which it is certainly the observable TCP position, relative to coordinate system at the base.

9.6. INVERSE KINEMATICS

Undoubtedly, it is a laborious problem obtains closed-form expressions for angles, depending on the position and orientation vectors. However, it is possible without much effort to determine the solution for some of them.

E.g., by dividing the equations 9.11 and 9.10 is easy to determine θ_1:

$$\boxed{\theta_1 = Tg^1\left(py / px\right)}$$

It is also easy to calculate θ_5 replacing the explicit value of $Cos(\theta_2 + \theta_3 + \theta_4)$ (equation 9.9) in equation 9.6:

$$\boxed{\theta_5 = -Seno^{-1}\left(sz / az\right)}$$

From here on you can appeal to computational tools to determine the remaining angles.

It is a relevant conclusion the following affirmation: the difficulty of obtaining inverse kinematic models is due to the need to isolate variables that are in the argument transcendental functions. In cases like this robot it is more difficult to obtain a solution, because all Transition-Homogeneous-Matrices have as variable parameters angles of revolute joints, so that the handling of sine and cosine functions must be continually addressed. In robots with prismatic joints this would be less difficult, but not negligible. In any case it is essential to understand that although they may be determined algebraic expressions for the inverse kinematic model, these expressions do not always will generate unique values and sometimes they will be nonexistent, for some positions and orientations which cannot be reached in practice.

CHAPTER 10

KINEMATICS OF LESS USED CONFIGURATIONS

"After climbing a great hill, one only finds that there are many more hills to climb".

Nelson Mandela

10.1. CONFIGURATION PPR

10.1.1. Forward Kinematics

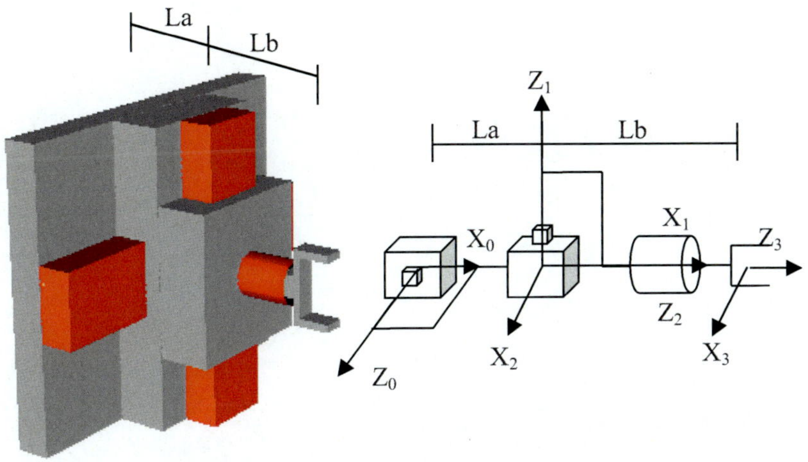

Figure 10.1 Functional representation and generalized coordinates systems of a PPR Robot

Table 10.1 DH Parameters of a PPR Robot

Joint	θ	d	a	α	q
1	0	q	La	-90	d1
2	-90	q	0	-90	d2
3	q	Lb	0	0	θ3

$$A_1^0 = \begin{bmatrix} 1 & 0 & 0 & La \\ 0 & 0 & 1 & 0 \\ 0 & -1 & 0 & d1 \\ 0 & 0 & 0 & 1 \end{bmatrix} \quad A_2^1 = \begin{bmatrix} 0 & 0 & 1 & 0 \\ -1 & 0 & 0 & 0 \\ 0 & -1 & 0 & d2 \\ 0 & 0 & 0 & 1 \end{bmatrix}$$

$$A_3^2 = \begin{bmatrix} C3 & -S3 & 0 & 0 \\ S3 & C3 & 0 & 0 \\ 0 & 0 & 1 & Lb \\ 0 & 0 & 0 & 1 \end{bmatrix}$$

$$T = \begin{bmatrix} 0 & 0 & 1 & La+Lb \\ -S3 & -C3 & 0 & d2 \\ C3 & -S3 & 0 & d1 \\ 0 & 0 & 0 & 1 \end{bmatrix}$$

10.1.2. Analysis of results

With no joints at secondary axes, this configuration can barely be considered a robot (really is more a two-axis mechanism). The value *px* can't be changed and the tool only can be moved changing its work direction (*slide*) in the YZ plane.

10.2. CONFIGURATION PRP

10.2.1. Forward Kinematics

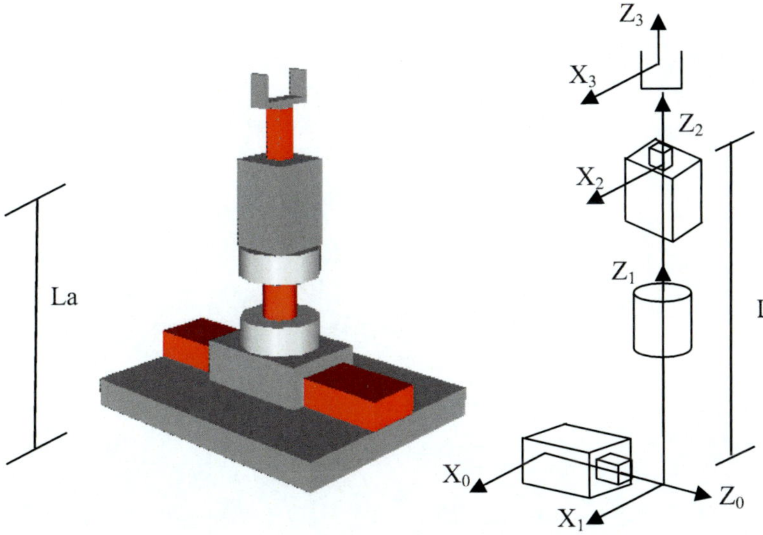

Figure 10.2 Functional representation and generalized coordinates systems of a PRP Robot

Table 10.2 DH Parameters of a PRP Robot

Joint	θ	d	a	α	q
1	0	q	0	90	d1
2	q	La	0	0	θ2
3	0	q	0	0	d3

$$A_1^0 = \begin{bmatrix} 1 & 0 & 0 & 0 \\ 0 & 0 & -1 & 0 \\ 0 & 1 & 0 & d1 \\ 0 & 0 & 0 & 1 \end{bmatrix} \quad A_2^1 = \begin{bmatrix} C2 & -S2 & 0 & 0 \\ S2 & C2 & 0 & 0 \\ 0 & 0 & 1 & La \\ 0 & 0 & 0 & 1 \end{bmatrix}$$

$$A_3^2 = \begin{bmatrix} 1 & 0 & 0 & 0 \\ 0 & 1 & 0 & 0 \\ 0 & 0 & 1 & d3 \\ 0 & 0 & 0 & 1 \end{bmatrix}$$

$$T = \begin{bmatrix} C2 & -S2 & 0 & 0 \\ 0 & 0 & -1 & -La-d3 \\ S2 & C2 & 0 & d1 \\ 0 & 0 & 0 & 1 \end{bmatrix}$$

10.2.2. Analysis of results

This configuration is even easier than before. In this case the x component of the position is constant and zero. No comments relevant to make.

10.3. CONFIGURATION PRR

Here a possible model PRR with three mutually perpendicular axes.

10.3.1 Forward Kinematics

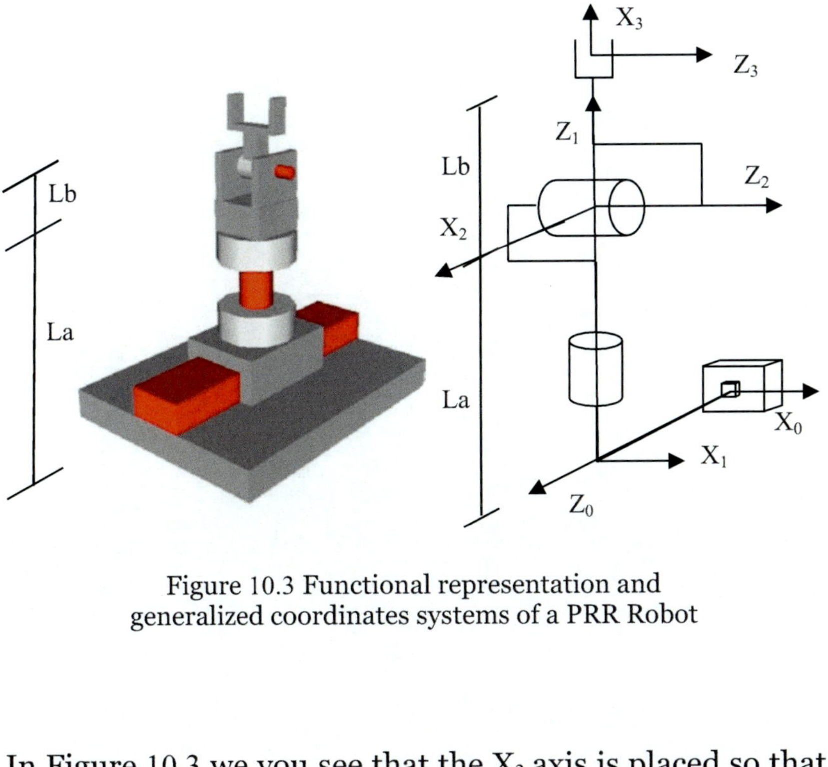

Figure 10.3 Functional representation and generalized coordinates systems of a PRR Robot

In Figure 10.3 we you see that the X_3 axis is placed so that the vector *approach* is not in the direction of the Z_3 axis as expected (this will have to consider it later to make a correct interpretation of the results obtained in matrix T).

Note that if X_3 copy the X_2 address, the dimension Lb would correspond to the Y_3 axis and would not appear in Table DH parameter.

Table 10.3 DH Parameters of PRR Robot

Joint	θ	d	a	α	q
1	0	q	0	-90	d1
2	q	La	0	-90	θ2
3	q	0	Lb	0	θ3

$$A_1^0=\begin{bmatrix}1 & 0 & 0 & 0\\ 0 & 0 & 1 & 0\\ 0 & -1 & 0 & d1\\ 0 & 0 & 0 & 1\end{bmatrix} \quad A_2^1=\begin{bmatrix}C2 & 0 & -S2 & 0\\ S2 & 0 & C2 & 0\\ 0 & -1 & 0 & La\\ 0 & 0 & 0 & 1\end{bmatrix}$$

$$A_3^2=\begin{bmatrix}C3 & -S3 & 0 & LbC3\\ S3 & C3 & 0 & LbS3\\ 0 & 0 & 1 & 0\\ 0 & 0 & 0 & 1\end{bmatrix}$$

$$T=\begin{bmatrix}C2C3 & -C2S3 & -S2 & LbC2C3\\ -S3 & C3 & 0 & La-LbS3\\ -S2C3 & S2S3 & -C2 & -LbS2C3+d1\\ 0 & 0 & 0 & 1\end{bmatrix}$$

In this special case the entries of submatrix of rotation correspond with the vectors: *normal, slide* and *approach*, as it can be observed in the following T matrix.

$$T = \begin{bmatrix} sx & ax & nx & px = LbC2C3 \\ sy & ay & ny & py = La - LbS3 \\ sz & az & nz & pz = d1 - LbS2C3 \\ 0 & 0 & 0 & 1 \end{bmatrix}$$

10.3.2. Analysis of results

Again, we can resort to notable amounts in joint parameters to evaluate the position equations. For example, in the condition shown in Figure 10.3 the parameters have the following values: $d_1 = 0$, $\theta_2 = -90$, $\theta_3 = -90$; additionally, is possible observe that $px = 0$, $py=0$, $pz = La+Lb$. As can be expected the evaluation of position equations reproduce these results.

$$px = Lb\mathrm{C}(-90)\mathrm{C}(-90) = 0$$
$$py = La - Lb\mathrm{S}(-90) = La + Lb$$
$$pz = 0 - Lb\mathrm{S}(-90)\mathrm{C}(-90) = 0$$

10.3.3. Inverse Kinematics

The simplicity of the robot does not merit further analysis. You can take the position equations of forward kinematics and isolate the joint parameters from them.

From the equation about *py*:

$$\theta_3 = Sin^{-1}(\frac{La - py}{Lb})$$

From here you have it:

$$C3 = \pm\frac{\sqrt{Lb^2 - (La - py)^2}}{Lb}$$

This expression is used in the *px* equation to get θ_2.

$$\theta_2 = Cos^{-1}(\pm\frac{px}{\sqrt{Lb^2 - (La - py)^2}})$$

Before you can determine d_1 it is necessary to find the expression of S_2:

$$S2 = \pm\sqrt{\frac{Lb^2 - (La - py)^2 - px^2}{Lb^2 - (La - py)^2}}$$

Finally, we can use S2 and C3 to obtain d_1 from the equation of *pz*:

$$d_1 = pz + \sqrt{Lb^2 - (La - py)^2 - px^2}$$

10.4. CONFIGURATION RPR

10.4.1. Forward Kinematics

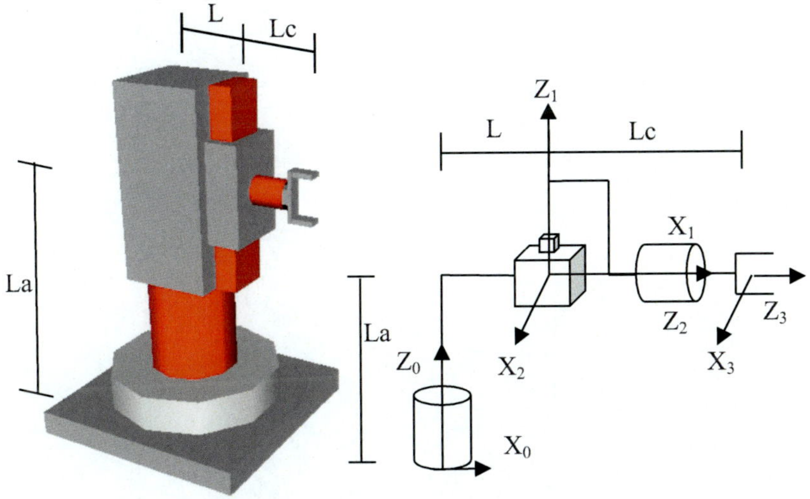

Figure 10.4 Functional representation and generalized coordinates systems of a RPR Robot

Table 10.4 DH Parameters of RPR Robot

Joint	θ	d	a	α	q
1	q	La	Lb	0	θ1
2	-90	q	0	-90	d2
3	q	Lc	0	0	θ3

$$A_1^0 = \begin{bmatrix} C1 & -S1 & 0 & LbC1 \\ S1 & C1 & 0 & LbS1 \\ 0 & 0 & 1 & La \\ 0 & 0 & 0 & 1 \end{bmatrix} \quad A_2^1 = \begin{bmatrix} 0 & 0 & 1 & 0 \\ -1 & 0 & 0 & 0 \\ 0 & -1 & 0 & d2 \\ 0 & 0 & 0 & 1 \end{bmatrix}$$

$$A_3^2 = \begin{bmatrix} C3 & -S3 & 0 & 0 \\ S3 & C3 & 0 & 0 \\ 0 & 0 & 1 & Lc \\ 0 & 0 & 0 & 1 \end{bmatrix}$$

$$T = \begin{bmatrix} S1C3 & -S1S3 & C1 & C1(Lb+Lc) \\ -C1C3 & C1S3 & S1 & S1(Lb+Lc) \\ -S3 & -C3 & 0 & d2+La \\ 0 & 0 & 0 & 1 \end{bmatrix}$$

10.4.2. Analysis of results

The results are completely anticipatable from a simple inspection. The "arm" Lb + Lc and parameter θ_1 clearly define the value of x and y components of the position, while the z component depends on the prismatic joint. The third joint influences only the working direction (slide) of the tool.

CHAPTER 11

KINEMATICS OF PUMA ROBOT

"It is always seems impossible until it's done".

Nelson Mandela

11.1. FUNCTIONAL REPRESENTATION AND COORDINATE SYSTEMS

Programmable Universal Machine for Assembly, o *Programmable Universal Manipulation Arm*. It is a 6R manipulator developed by Universal Automation (Unimation) in 1978, with the support of General Motors and based on designs Robotics pioneer Victor Scheinman (the creator of the Stanford Robot).

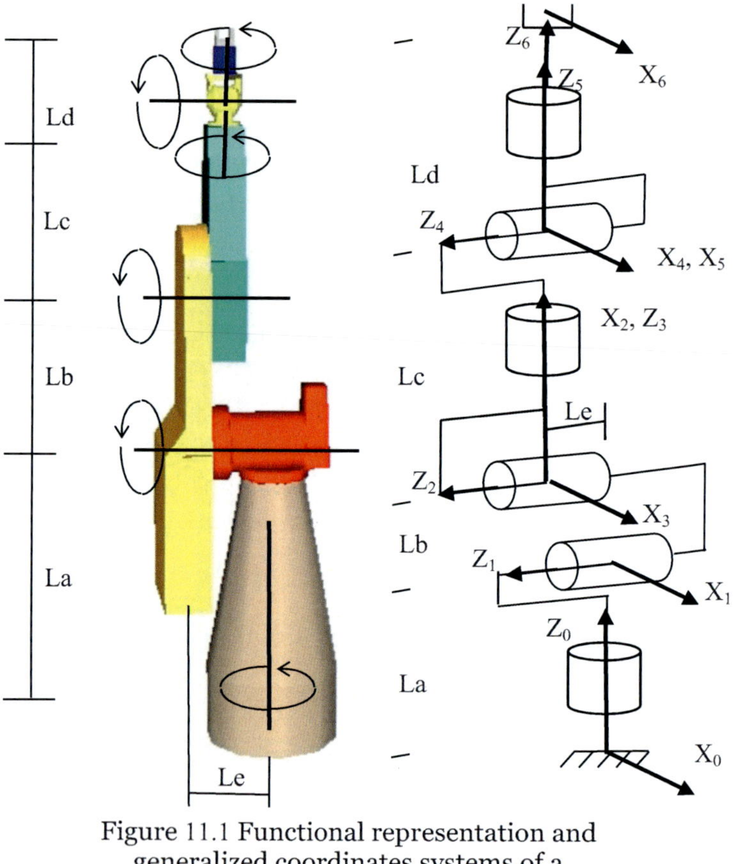

Figure 11.1 Functional representation and generalized coordinates systems of a PUMA Robot; and virtual model from software RoboWorks Demo 2.0

11.2. TABLE OF DH PARAMETERS

Table 11.1 DH Parameters of PUMA Robot

Joint	θ	d	a	α	q
1	q_1	La	0	90	θ_1
2	q_2	Le	Lb	0	θ_2
3	q_3	0	0	-90	θ_3
4	q_4	Lc	0	90	θ_4
5	q_5	0	0	-90	θ_5
6	q_6	Ld	0	0	θ_6

11.3. TRANSITION HOMOGENEOUS MATRICES

General equation:

$$^{i-1}A_i = \begin{bmatrix} C\theta & -C\alpha S\theta & S\alpha S\theta & aC\theta \\ S\theta & C\alpha C\theta & -S\alpha C\theta & aS\theta \\ 0 & S\alpha & C\alpha & d \\ 0 & 0 & 0 & 1 \end{bmatrix}$$

Particular matrices:

$$A_1^0 = \begin{bmatrix} C1 & 0 & S1 & 0 \\ S1 & 0 & -C1 & 0 \\ 0 & 1 & 0 & La \\ 0 & 0 & 0 & 1 \end{bmatrix} \quad A_2^1 = \begin{bmatrix} C2 & -S2 & 0 & LbC2 \\ S2 & C2 & 0 & LbS2 \\ 0 & 0 & 1 & Le \\ 0 & 0 & 0 & 1 \end{bmatrix}$$

$$A_3^2 = \begin{bmatrix} C3 & 0 & -S3 & 0 \\ S3 & 0 & C3 & 0 \\ 0 & -1 & 0 & 0 \\ 0 & 0 & 0 & 1 \end{bmatrix} \quad A_4^3 = \begin{bmatrix} C4 & 0 & S4 & 0 \\ S4 & 0 & -C4 & 0 \\ 0 & 1 & 0 & Lc \\ 0 & 0 & 0 & 1 \end{bmatrix}$$

$$A_5^4 = \begin{bmatrix} C5 & 0 & -S5 & 0 \\ S5 & 0 & C5 & 0 \\ 0 & -1 & 0 & 0 \\ 0 & 0 & 0 & 1 \end{bmatrix} \quad A_5^6 = \begin{bmatrix} C6 & -S6 & 0 & 0 \\ S6 & C6 & 0 & 0 \\ 0 & 0 & 1 & Ld \\ 0 & 0 & 0 & 1 \end{bmatrix}$$

11.4. ENTRIES OF TRANSFORMATIOM MATRIX OF THE ROBOT

As in the case of robot 5R, from the product of Transition Homogeneous Matrices, the matrix T is obtained but, because of their size, below only its elements are shown.

$$nx = S6\left(S4(C1C2C3 - C1S2S3) - C4S1\right) + C6\begin{pmatrix} C5(S1S4 + C4(C1C2C3 - C1S2S3)) \\ - S5(C1C2S3 + C1C3S2) \end{pmatrix}$$

$$ny = C6\left(C5(C4(C2C3S1 - S1S2S3) - C1S4) - S5(C2S1S3 + C3S1S2)\right)$$

$$+ S6\left(S4(C2C3S1 - S1S2S3) + C1C4\right)$$

$$nz = S4S6\left(C2S3 + C3S2\right) - C6\left(S5(S2S3 - C2C3) - C4C5(C2S3 + C3S2)\right)$$

$$sx = C6\left(S4(C1C2C3 - C1S2S3) - C4S1\right) - S6\begin{pmatrix} C5(S1S4 + C4(C1C2C3 - C1S2S3)) \\ - S5(C1C2S3 + C1C3S2) \end{pmatrix}$$

$$sy = C6\left(S4(C2C3S1 - S1S2S3) + C1C4\right) - S6\begin{pmatrix} C5(C4(C2C3S1 - S1S2S3) - C1S4) \\ - S5(C2S1S3 + C3S1S2) \end{pmatrix}$$

$$sz = S6\left(S5(S2S3 - C2C3) - C4C5(C2S3 + C3S2)\right) + C6S4\left(C2S3 + C3S2\right)$$

$$ax = - C5(C1C2S3 + C1C3S2) - S5(S1S4 + C4(C1C2C3 - C1S2S3))$$

$$ay = - C5(C2S1S3 + C3S1S2) - S5(C4(C2C3S1 - S1S2S3) - C1S4)$$

$$az = - C5(S2S3 - C2C3) - C4S5(C2S3 + C3S2)$$

$$px = LeS1 - Lc\left(C1C2S3 + C1C3S2\right) - Ld\begin{pmatrix} C5(C1C2S3 + C1C3S2) \\ + S5(S1S4 + C4(C1C2C3 - C1S2S3)) \end{pmatrix}$$

$$+ C1C2Lb$$

$$py = C2LbS1 - Lc\left(C2S1S3 + C3S1S2\right) - C1Le$$

$$- Ld\left(C5(C2S1S3 + C3S1S2) + S5(C4(C2C3S1 - S1S2S3) - C1S4)\right)$$

$$pz = La - Ld(C5(S2S3 - C2C3) + C4S5(C2S3 + C3S2)) - Lc(S2S3 - C2C3) + LbS2$$

A quick check of the validity of this result can be obtained by evaluating the position equations: $\theta_1 = 0^o$, $\theta_2 = 90^o$, $\theta_3 = -90^o$, $\theta_4 = 0^o$, $\theta_5 = 0^o$, $\theta_6 = 0^o$. With these values, taken from the condition shown to the robot in figure 11.1, it is confirmed that $px = 0$, $py = -Le$, $pz = La + Lb + Lc + Ld$, which is the TCP position to fully upright Manipulator, as it is seen in that figure.

CHAPTER 12

KINEMATICS OF STANFORD ARM

"The best prophet of the future is the past".

George Gordon Byron

"Study the past if you would divine the future".

Kung Fu Tse

Designed in 1969 at the Stanford Artificial Intelligence Laboratory by Victor Scheinman, this mythical manipulator has 6 degrees of freedom RRPRRR type and has earned its place in the history of industrial robotics for being the first to have a fully computerized control system and display its versatility in varied applications where a computer vision system was required.

Figure 12.1 Stanford Arm

12.1. FUNCTIONAL REPRESENTATION AND COORDINATE SYSTEMS

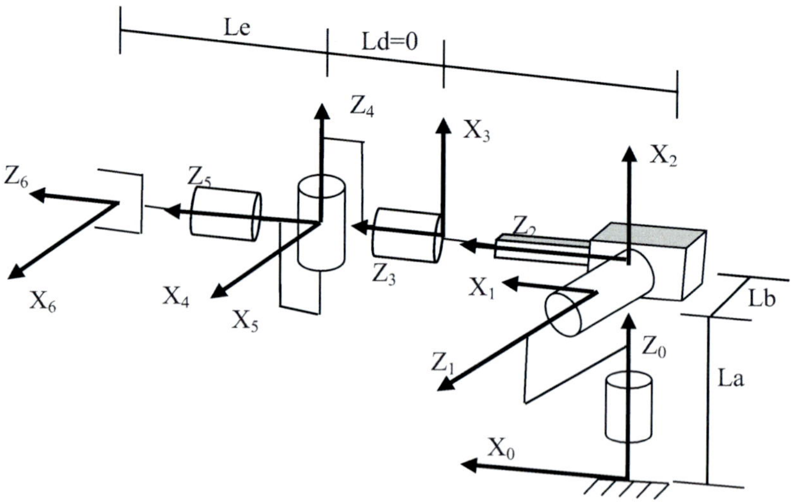

Figure 12.2 Functional representation and generalized coordinates systems of STANFORD

12.2. TABLE OF DH PARAMETERS

Table 12.1 DH Parameters of STANFORD

Joint	θ	d	a	α	q
1	q_1	La	0	-90	θ_1
2	q_2	-Lb	0	-90	θ_2
3	0	q_3	0	0	d_3
4	q_4	Ld=0	0	-90	θ_4
5	q_5	0	0	90	θ_5
6	q_6	Le	0	0	θ_6

The configuration of the wrist (joints 4, 5 and 6) is of spherical type, with a coincidence of the origins of the S_3, S_4 and S_5; and a null distance Ld.

12.3. TRANSITION HOMOGENEOUS MATRICES

General equation:

$$^{i-1}A_i = \begin{bmatrix} C\theta & -C\alpha S\theta & S\alpha S\theta & aC\theta \\ S\theta & C\alpha C\theta & -S\alpha C\theta & aS\theta \\ 0 & S\alpha & C\alpha & d \\ 0 & 0 & 0 & 1 \end{bmatrix}$$

Particular matrices:

$$A_1^0 = \begin{bmatrix} C1 & 0 & -S1 & 0 \\ S1 & 0 & C1 & 0 \\ 0 & -1 & 0 & La \\ 0 & 0 & 0 & 1 \end{bmatrix} \quad A_2^1 = \begin{bmatrix} C2 & 0 & -S2 & 0 \\ S2 & 0 & C2 & 0 \\ 0 & -1 & 0 & -Lb \\ 0 & 0 & 0 & 1 \end{bmatrix}$$

$$A_3^2 = \begin{bmatrix} 1 & 0 & 0 & 0 \\ 0 & 1 & 0 & 0 \\ 0 & 0 & 1 & d3 \\ 0 & 0 & 0 & 1 \end{bmatrix} \quad A_4^3 = \begin{bmatrix} C4 & 0 & -S4 & 0 \\ S4 & 0 & C4 & 0 \\ 0 & -1 & 0 & 0 \\ 0 & 0 & 0 & 1 \end{bmatrix}$$

$$A_5^4 = \begin{bmatrix} C5 & 0 & S5 & 0 \\ S5 & 0 & -C5 & 0 \\ 0 & 1 & 0 & 0 \\ 0 & 0 & 0 & 1 \end{bmatrix} \quad A_6^5 = \begin{bmatrix} C6 & -S6 & 0 & 0 \\ S6 & C6 & 0 & 0 \\ 0 & 0 & 1 & Le \\ 0 & 0 & 0 & 1 \end{bmatrix}$$

12.4. TRANSFORMATION MATRIX OF THE ROBOT

$$T = A_0^1 * A_1^2 * A_2^3 * A_3^4 * A_4^5 * A_5^6$$

$$T = \begin{bmatrix} nx & sx & ax & px \\ ny & sy & ay & py \\ nz & sz & az & pz \\ 0 & 0 & 0 & 1 \end{bmatrix}$$

nx = ((C1C2C4+S1S4)C5+C1S2S5)C6+(-C1C2S4+S1C4)S6
ny = ((S1C2C4-C1S4)C5+S1S2S5)C6+(-S1C2S4-C1C4)S6
nz = (-S2C4C5+C2S5)C6+S2S4S6

sx = -((C1C2C4+S1S4)C5+C1S2S5)S6+(-C1C2S4+S1C4)C6
sy = -((S1C2C4-C1S4)C5+S1S2S5)S6+(-S1C2S4-C1C4)C6
sz = -(-S2C4C5+C2S5)S6+S2S4C6

ax = (C1C2C4+S1S4)S5-C1S2C5
ay = (S1C2C4-C1S4)S5-S1S2C5
az = -S2C4S5-C2C5

px = ((C1C2C4+S1S4)S5-C1S2C5)Le-C1S2d3+S1Lb
py = ((S1C2C4-C1S4)S5-S1S2C5)Le-S1S2d3-C1Lb
pz = (-S2C4S5-C2C5)Le-C2d3+La

12.5. ANALYSIS OF RESULTS

Looking carefully the functional representation of the figure 12.2, we find the robot in a location produced by the following values of joint parameters: $\theta_1 = 0^\circ$; $\theta_2 = -90^\circ$, $\theta_4 = -90^\circ$, $\theta_5 = 0^\circ$, $\theta_6 = 0^\circ$. For this set of values is clear that: $px = d_3 + Le$, $py = -Lb$, $pz = La$.

These same results are obtained by evaluating the expressions in the fourth column of matrix T, confirming the validity of the model.

CHAPTER 13

KINEMATICS OF PRRPRR ROBOT

"Inspiration exists, but it has to find you working".

Pablo Picasso

The robot then studied is a simplified model of a design created by Professor Panfilino Mónaco in the Experimental Institute of Technology La Victoria, Aragua, Venezuela. The robot has a seventh joint, which is not considered herein.

13.1. FUNCTIONAL REPRESENTATION AND COORDINATE SYSTEMS

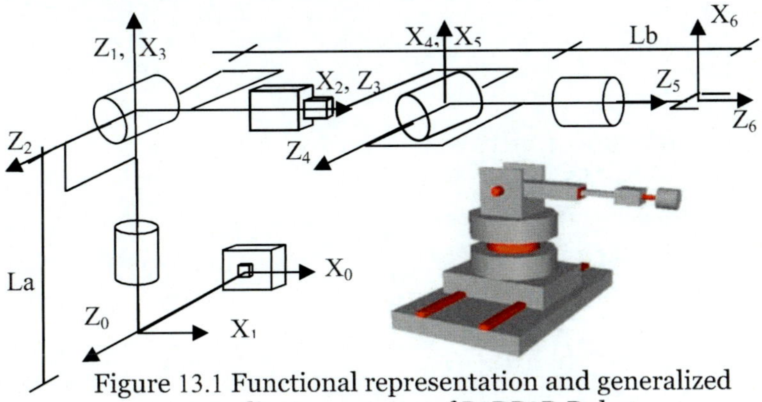

Figure 13.1 Functional representation and generalized coordinates systems of P2RP2R Robot

13.2. TABLE OF DH PARAMETERS

Table 13.1. DH Parameters P2RP2R Robot

Joint	θ	d	a	α	q
1	0	q	0	-90	d1
2	q	La	0	90	θ2
3	q	0	0	90	θ3
4	0	q	0	-90	d4
5	q	0	0	90	θ5
6	q	Lb	0	0	θ6

13.3. TRANSITION HOMOGENEOUS MATRICES

General equation:

$$
{}^{i-1}A_i = \begin{bmatrix} C\theta & -C\alpha S\theta & S\alpha S\theta & aC\theta \\ S\theta & C\alpha C\theta & -S\alpha C\theta & aS\theta \\ 0 & S\alpha & C\alpha & d \\ 0 & 0 & 0 & 1 \end{bmatrix}
$$

Particular matrices:

$$
A_1^0 = \begin{bmatrix} 1 & 0 & 0 & 0 \\ 0 & 0 & 1 & 0 \\ 0 & -1 & 0 & d1 \\ 0 & 0 & 0 & 1 \end{bmatrix} \quad A_2^1 = \begin{bmatrix} C2 & 0 & S2 & 0 \\ S2 & 0 & -C2 & 0 \\ 0 & 1 & 0 & La \\ 0 & 0 & 0 & 1 \end{bmatrix}
$$

$$A_3^2 = \begin{bmatrix} C3 & 0 & S3 & 0 \\ S3 & 0 & -C3 & 0 \\ 0 & 1 & 0 & 0 \\ 0 & 0 & 0 & 1 \end{bmatrix} \quad A_4^3 = \begin{bmatrix} 1 & 0 & 0 & 0 \\ 0 & 0 & 1 & 0 \\ 0 & -1 & 0 & d4 \\ 0 & 0 & 0 & 1 \end{bmatrix}$$

$$A_5^4 = \begin{bmatrix} C5 & 0 & S5 & 0 \\ S5 & 0 & -C5 & 0 \\ 0 & 1 & 0 & 0 \\ 0 & 0 & 0 & 1 \end{bmatrix} \quad A_6^5 = \begin{bmatrix} C6 & -S6 & 0 & 0 \\ S6 & C6 & 0 & 0 \\ 0 & 0 & 1 & Lb \\ 0 & 0 & 0 & 1 \end{bmatrix}$$

13.4. ENTRIES OF THE TRANSFORMATION MATRIX OF THE ROBOT

The elements of the matrix T obtained by post multiplication of the above matrices are:

nx = (C2C3C5-C2S3S5)C6+S2S6

ny = (S3C5+C3S5)C6

nz = (-S2C3C5+S2S3S5)C6+C2S6

sx = -(C2C3C5-C2S3S5)S6+S2C6

sy = -(S3C5+C3S5)S6

sz = -(-S2C3C5+S2S3S5)S6+C2C6

ax = C2C3S5+C2S3C5

ay = S3S5-C3C5

az = -S2C3S5-S2S3C5

px = (C2C3S5+C2S3C5)Lb+C2S3d4

py = (S3S5-C3C5)Lb-C3d4+La

pz = (-S2C3S5-S2S3C5)Lb-S2S3d4+d1

13.5. ANALYSIS OF RESULTS

A simple check on previous results may be performed by evaluating position equations with the combination of joints parameters which reproduce the condition shown in the figure corresponding to the functional representation. Such a combination is: $d_1=0$, $\theta_2=0$, $\theta_3=90$, $d_4=0$, $\theta_5=0$, $\theta_6=0$.

In the figure 13.1 it can see that: *px =Lb*, *py = La*, *pz = 0*.

px = (C(0)C(90)S(0)+C(0)S(90)C(0))Lb+C(0)S(90)*0=Lb
py = (S(90)S(0)-C(90)C(0))Lb-C(90)*0+La=La
pz = (-S(0)C(90)S(0)-S(0)S(90)C(0))Lb-S(0)S(90)*0+0=0

If an analysis like this is repeated for the same parameter values except θ_3, which is set at 180°, it produces the condition shown in figure 13.2, where: *px* = 0, *py* = *La* + *Lb*, *pz* = 0.

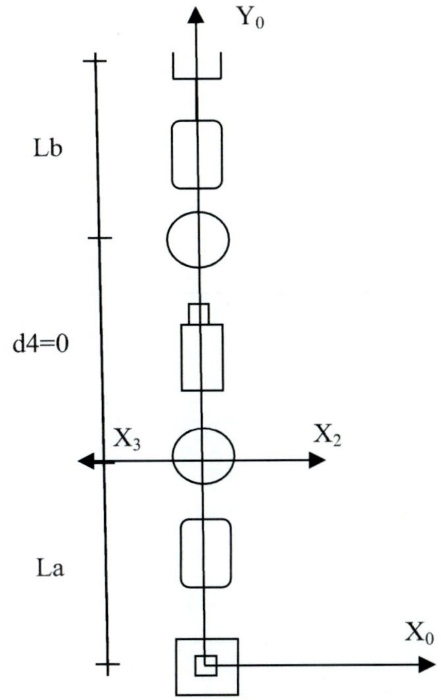

Figure 13.2 XY plane of the P2RP2R robot with $d_1 = 0$, $\theta_2 = 0$, $\theta_3 = 180^{\circ}$, $d_4 = 0$, $\theta_5 = 0$, $\theta_6 = 0$

px = (C(0)C(180)S(0)+C(0)S(180)C(0))Lb+C(0)S(180)*0=0

py = (S(180)S(0)-C(180)C(0))Lb-C(180)*0+La=La+Lb

pz = (-S(0)C(180)S(0)-S(0)S(180)C(0))Lb-S(0)S(180)*0+0=0

CHAPTER 14

KINEMATICS OF ANTHROPOMORPHIC 6R ROBOT

"Start by doing what's necessary; then do what's possible; and suddenly you are doing the impossible".

Giovanni di Pietro di Bernardone

"Judge your success by what you had to give up in order to get it".

Tenzin Gyatso

The most popular robot used in the Industry is the anthropomorphic of six axes. Almost all manufacturers of Industrial Robots have one or more series of these, based on the load capacity and/or type of application areas. As example, in this chapter it is studied the robot KR 180R-3200 PA, built by KUKA Roboter GmbH, an important manufacturer of robots and automated solutions for the Industry.

Figure 14.1 KUKA Robot KR 180R-3200 PA
Courtesy of: Neumática Rotonda, Valencia, Venezuela

14.1. FUNCTIONAL REPRESENTATION AND COORDINATE SYSTEMS

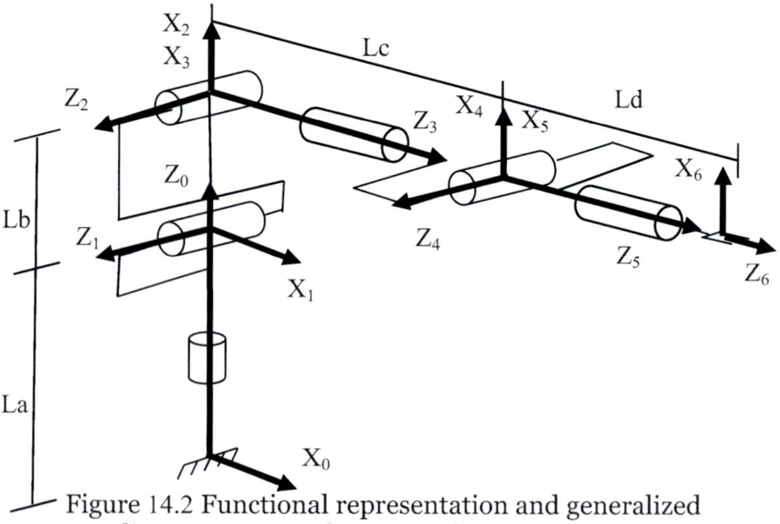

Figure 14.2 Functional representation and generalized coordinates systems of KUKA Robot KR 180R-3200 PA

14.2. TABLE OF DH PARAMETERS

Table 14.1 DH Parameters of Robot KUKA Robot KR 180R-3200 PA

Joint	θ	d	a	α	q
1	q_1	La	0	90	θ_1
2	q_2	0	Lb	0	θ_2
3	q_3	0	0	90	θ_3
4	q_4	Lc	0	-90	θ_4
5	q_5	0	0	90	θ_5
6	q_6	Ld	0	0	θ_6

14.3. TRANSITION HOMOGENEOUS MATRICES

General equation:

$$^{i-1}A_i = \begin{bmatrix} C\theta & -C\alpha S\theta & S\alpha S\theta & aC\theta \\ S\theta & C\alpha C\theta & -S\alpha C\theta & aS\theta \\ 0 & S\alpha & C\alpha & d \\ 0 & 0 & 0 & 1 \end{bmatrix}$$

Particular matrices:

$$A_1^0 = \begin{bmatrix} C1 & 0 & S1 & 0 \\ S1 & 0 & -C1 & 0 \\ 0 & 1 & 0 & La \\ 0 & 0 & 0 & 1 \end{bmatrix} \quad A_2^1 = \begin{bmatrix} C2 & -S2 & 0 & LbC2 \\ S2 & C2 & 0 & LbS2 \\ 0 & 0 & 1 & 0 \\ 0 & 0 & 0 & 1 \end{bmatrix}$$

$$A_3^2 = \begin{bmatrix} C3 & 0 & S3 & 0 \\ S3 & 0 & -C3 & 0 \\ 0 & 1 & 0 & 0 \\ 0 & 0 & 0 & 1 \end{bmatrix} \quad A_4^3 = \begin{bmatrix} C4 & 0 & -S4 & 0 \\ S4 & 0 & C4 & 0 \\ 0 & -1 & 0 & Lc \\ 0 & 0 & 0 & 1 \end{bmatrix}$$

$$A_5^4 = \begin{bmatrix} C5 & 0 & S5 & 0 \\ S5 & 0 & -C5 & 0 \\ 0 & 1 & 0 & 0 \\ 0 & 0 & 0 & 1 \end{bmatrix} \quad A_6^5 = \begin{bmatrix} C6 & -S6 & 0 & 0 \\ S6 & C6 & 0 & 0 \\ 0 & 0 & 1 & Ld \\ 0 & 0 & 0 & 1 \end{bmatrix}$$

14.4. ENTRIES OF TRANSFORMATIOM MATRIX OF THE ROBOT

$nx =$ (C1C2C3-C1S2S3)(C4C5C6-S4S6)+S1(S4C5C6+C4S6)

-(C1C2S3+C1S2C3)S5C6

$ny =$ (S1C2C3-S1S2S3)(C4C5C6-S4S6)-C1(S4C5C6+C4S6)

-(S1C2S3+S1S2C3)S5C6

$nz =$ (S2C3+C2S3)(C4C5C6-S4S6)-(S2S3-C2C3)S5C6

$sx = -$(C1C2C3-C1S2S3)(C4C5S6+S4C6)+S1(-S4C5S6+C4C6)

+(C1C2S3+C1S2C3)S5S6

$sy = -$(S1C2C3-S1S2S3)(C4C5S6+S4C6)-C1(-S4C5S6+C4C6)

+(S1C2S3+S1S2C3)S5S6

$sz = -$(S2C3+C2S3)(C4C5S6+S4C6)+(S2S3-C2C3)S5S6

$ax =$ (C1C2C3-C1S2S3)C4S5+S1S4S5+(C1C2S3+C1S2C3)C5

$ay =$ (S1C2C3-S1S2S3)C4S5-C1S4S5+(S1C2S3+S1S2C3)C5

$az =$ (S2C3+C2S3)C4S5+(S2S3-C2C3)C5

px = (C1C2C3-C1S2S3)LdC4S5+S1LdS4S5

+(C1C2S3+C1S2C3)(LdC5+Lc)+LbC1C2

py = (S1C2C3-S1S2S3)LdC4S5-C1LdS4S5

+(S1C2S3+S1S2C3)(LdC5+Lc)+LbS1C2

pz = (S2C3+C2S3)LdC4S5+(S2S3-C2C3)(LdC5+Lc)+LbS2+La

14.5. ANALYSIS OF RESULTS

Observing the functional representation of the figure 14.2 the robot is found in a condition which reached thanks to the following joint parameters: $\theta_1 = 0°$; $\theta_2 = 90°$, $\theta_3 = 0°$, $\theta_4 = 0°$, $\theta_5 = 0°$, $\theta_6 = 0°$. For this set of values, it is clear that: $px = Lc+Ld$, $py = 0$, $pz = La+Lb$.

The same results are obtained evaluating the equations in the last column of the T matrix, which is reason to consider the validity of the model.

As exercise, if it will change the value of θ_3 to 180°, keeping all another angles, it is logical that $px = -Lc-Ld$, $py = 0$, $pz = La+Lb$. The new evaluation of the position equations is easy and allows confirming the hoped results.

CHAPTER 15

THE FIRST INDUSTRIAL ROBOT: UNIMATE

"There was nowhere to go to get information, so I generated it ".

George Devol

This is recognized as the first Industrial robot. It was created by George Devol Jr., who introduced his patent in 1954. Later, associated with Joseph Engelberger, Devol founded the company Unimation (Universal Automation), where it was built the original UNIMATE. The robot was used to work with heated die-casting machines, at the Inland Fisher Guide Plant of General Motors, in West Trenton (New Jersey), in 1961.

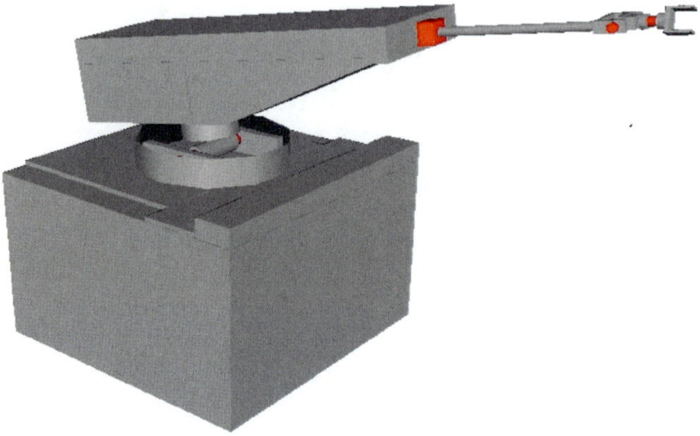

Figure 15.1 The Robot UNIMATE

The Unimate has 5 Degrees of Freedom: the axes 1, 2 and 3 are RRP, corresponding to a spherical configuration; the 4th axis implements the *Pitch* angle and the 5th axis, the *Roll* angle.

15.1. FUNCTIONAL REPRESENTATION AND COORDINATE SYSTEMS

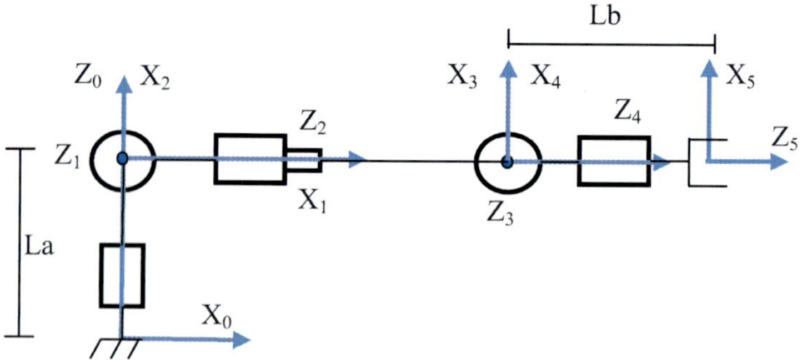

Figure 15.2 Functional representation and generalized coordinates systems of Robot UNIMATE

15.2. TABLE OF DH PARAMETERS

Table 15.1 DH Parameter of Robot UNIMATE

Joint	θ	d	a	α	q
1	q_1	La	0	90	θ_1
2	q_2	0	0	90	θ_2
3	0	q_3	0	-90	d_3
4	q_4	0	0	90	θ_4
5	q_5	Lb	0	0	θ_5

15.3. TRANSITION HOMOGENEOUS MATRICES

General equation:

$$ {}^{i-1}A_i = \begin{bmatrix} C\theta & -C\alpha S\theta & S\alpha S\theta & aC\theta \\ S\theta & C\alpha C\theta & -S\alpha C\theta & aS\theta \\ 0 & S\alpha & C\alpha & d \\ 0 & 0 & 0 & 1 \end{bmatrix} $$

Particular matrices

$$ A_1^0 = \begin{bmatrix} C1 & 0 & S1 & 0 \\ S1 & 0 & -C1 & 0 \\ 0 & 1 & 0 & La \\ 0 & 0 & 0 & 1 \end{bmatrix} \quad A_2^1 = \begin{bmatrix} C2 & 0 & S2 & 0 \\ S2 & 0 & -C2 & 0 \\ 0 & 1 & 0 & 0 \\ 0 & 0 & 0 & 1 \end{bmatrix} $$

$$A_3^2 = \begin{bmatrix} 1 & 0 & 0 & 0 \\ 0 & 0 & 1 & 0 \\ 0 & -1 & 0 & d3 \\ 0 & 0 & 0 & 1 \end{bmatrix} \quad A_4^3 = \begin{bmatrix} C4 & 0 & S4 & 0 \\ S4 & 0 & -C4 & 0 \\ 0 & 1 & 0 & 0 \\ 0 & 0 & 0 & 1 \end{bmatrix}$$

$$A_5^4 = \begin{bmatrix} C5 & -S5 & 0 & 0 \\ S5 & C5 & 0 & 0 \\ 0 & 0 & 1 & Lb \\ 0 & 0 & 0 & 1 \end{bmatrix}$$

15.4. ENTRIES OF TRANSFORMATIOM MATRIX OF THE ROBOT

nx $= S1C2C4C5 - C1S2S4C5 + S1S5$

ny $= S1C2C4C5 - S1S2S4C5 - C1S5$

nz $= S2C4C5 + C2S4C5$

sx $= -C1C2C4S5 + C1S2S4S5 + S1C5$

sy $= -S1C2C4S5 + S1S2S4S5 - C1C5$

sz $= -S2C4S5 - C2S4S5$

ax $= C1C2S4 + C1S2C4$

ay $= S1C2S4 + S1S2C4$

az $= S2S4 - C2C4$

px $= LbC1C2S4 + LbC1S2C4 + d3C1S2$

py $= LbS1C2S4 + LbS1S2C4 + d3S1S2$

pz $= LbS2S4 - LbC2C4 - \mathrm{d3C2} + \mathrm{La}$

15.5. ANALYSIS OF RESULTS

The position of TCP in the figure 15.2 (*px* = Lb, *py* = 0, *pz* = La) is according with following values: θ_1= 0º, θ_2 = 90º, d_3 = 0, θ_4 = 0º, θ_5 = 0º. The same results are obtained evaluating the position equations, which is positive evidence about the model is correct.

$$\text{px} = LbC(0)C(90)S(0) + LbC(0)S(90)C(0) + (0)C(0)S(90) = Lb$$

$$\text{py} = LbS(0)C(90)S(0) + LbS(0)S(90)C(0) + (0)S(0)S(90) = 0$$

$$\text{pz} = LbS(90)S(0) - LbC(90)C(0) - (0)C(90) + La = La$$

APENDIX A

SCRIPTS OF USED FUNCTIONS

```
% ROTX      Rotation around X axis
%
%     ROTX(alpha)    Returns    the    Homogeneous
transformation
%     of rotation around X axis, an angle alpha.
%
%     T  =  rotx(alpha)  assigns  the  result  (4x4
matrix ) to a variable T
%
%     See too ROTY, ROTZ.
%
%  ************************************************
%  * Developed independently by Wilmer Sanz F. in *
%  * Dec. 2002, but preceded by a function of the *
%  * same name created in 1990 by Peter Corke.    *
%  * Differs from Corke's version in that the     *
%  * argument must be entered in degrees,         *
%  * not radians.                                 *
%  ************************************************

function r = rotx(alpha)
      % Conversion de grados a radianes
  a=alpha*pi/180;
      %Conversion of degrees to radians
  ca = cos(a);
      sa = sin(a);
      r = [1      0      0      0
            0      ca     -sa    0
            0      sa     ca     0
            0      0      0      1];
```

```
% ROTY       Rotation around Y axis
%
%      ROTY(phi)     Returns     the     Homogeneous
transformation
%      of rotation around Y axis, an angle phi.
%
%      T = roty(phi) assigns the result (4x4 matrix
) to a variable T
%
%      See too ROTX, ROTZ.
%
%   ***********************************************
%   * Developed independently by Wilmer Sanz F. in *
%   * Dec. 2002, but preceded by a function of the *
%   * same name created in 1990 by Peter Corke.    *
%   * Differs from Corke's version in that the     *
%   * argument must be entered in degrees,         *
%   * not radians.                                 *
%   ***********************************************

function r = roty(phi)
  % Conversion de grados a radianes
      f=phi*pi/180;
  %Conversion of degrees to radians
      cf = cos(f);
      sf = sin(f);
      r = [cf     0     sf     0
            0     1     0      0
          -sf     0     cf     0
            0     0     0      1];
```

```
% ROTZ      Rotation around Z axis
%
%     ROTZ(theta)    Returns    the    Homogeneous
transformation
%     of rotation around Z axis, an angle theta.
%
%     T  =  rotz(theta)  assigns  the  result  (4x4
matrix ) to a variable T
%
%     See too ROTX, ROTY.
%
%  ************************************************
%  * Developed independently by Wilmer Sanz F. in *
%  * Dec. 2002, but preceded by a function of the *
%  * same name created in 1990 by Peter Corke.    *
%  * Differs from Corke's version in that the     *
%  * argument must be entered in degrees,         *
%  * not radians.                                 *
%  ************************************************

function r = rotz(t)
  % Conversion de grados a radianes
      t=t*pi/180;
  % Conversion of degrees to radians
      ct = cos(t);
      st = sin(t);
      r = [ct     -st   0     0
              st  ct    0     0
              0   0     1     0
              0   0     0     1];
```

```
% ROTXTRANS Rotation  around  X  axis,  followed  by
translation
%
%         ROTXTRANS(alpha,px,py,pz)   Returns   the
Homogeneous transformation
%    of rotation around X axis, an angle alpha,
followed by translation
%   given in a vector p.
%
%   T = rotxtrans(alpha,px,py,pz) assigns the
result (4x4 matrix ) to a
%   variable T
%   T = rotxtrans(alpha,p) same result but requires
p as row vector (size
%   1x3)
%
%    See   too   ROTYTRANS,   ROTZTRANS,   TRANSROTX,
TRANSROTY, TRANSROTZ.
%
%  ***********************************************
%  * Developed by Wilmer Sanz F. Dec 2002         *
%  * The angle must be entered in degrees         *
%  ***********************************************
function r = rotxtrans(alpha, px, py, pz)
    % Conversion de grados a radianes
            a=alpha*pi/180;
    % Conversion of degrees to radians
            ca = cos(a);
            sa = sin(a);
if nargin == 4,
      r = [1     0     0     px
      0     ca    -sa   py
      0     sa    ca    pz
      0     0     0     1];
elseif nargin == 2,
      r = [1     0     0     px(1)
      0     ca    -sa   px(2)
      0     sa    ca    px(3)
      0     0     0     1];
end
```

```
% ROTYTRANS Rotation around Y axis, followed by
translation
%
%     ROTYTRANS(phi,px,py,pz)      Returns      the
Homogeneous transformation
%     of  rotation  around  Y  axis,  an  angle  phi,
followed by translation
%     given in a vector p.
%
%   T = rotytrans(phi,px,py,pz) assigns the result
(4x4 matrix ) to a
%  variable T
%  T = rotytrans(phi,p) same result but requires p
as row vector (size
%  1x3)
%
%     See   too   ROTXTRANS,   ROTZTRASN,   TRANSROTX,
TRANSROTY, TRANSROTZ.
%
%  ************************************************
%  * Developed by Wilmer Sanz F. Dec 2002          *
%  * The angle must be entered in degrees          *
%  ************************************************
function r = rotytrans(phi, px, py, pz)
    % Conversion de grados a radianes
            f=phi*pi/180;
    % Conversion of degrees to radians
            cf = cos(f);
            sf = sin(f);
if nargin == 4,
   r = [cf  0     sf    px
              0   1     0     py
            -sf   0     cf    pz
              0   0     0     1];
elseif nargin == 2,
   r = [cf  0     sf    px(1)
        0   1     0     px(2)
            -sf   0     cf    px(3)
              0   0     0     1];
end
```

```
% ROTZTRANS Rotation  around  Z  axis,  followed  by
translation
%
%     ROTZTRANS(theta,px,py,pz)     Returns     the
Homogeneous transformation
%     of rotation around Z axis, an angle theta,
followed by translation
%     given in a vector p.
%
%  T = rotztrans(theta,px,py,pz) assigns the result
(4x4 matrix ) to a
%  variable T
%  T = rotztrans(theta,p) same result but requires
p as row vector (size
%  1x3)
%
%     See   too   ROTXTRANS,   ROTYTRASN,   TRANSROTX,
TRANSROTY, TRANSROTZ.
%
%  ************************************************
%  * Developed by Wilmer Sanz F. Dec 2002          *
%  * The angle must be entered in degrees          *
%  ************************************************
function r = rotztrans(theta, px, py, pz)
    % Conversion de grados a radianes
            t=theta*pi/180;
    % Conversion of degrees to radians
            ct = cos(t);
            st = sin(t);
if nargin == 4,
   r = [ct -st    0     px
        st  ct    0     py
              0   0     1     pz
              0   0     0     1];
elseif nargin == 2,
   r = [ct  -st   0     px(1)
              st  ct    0     px(2)
              0   0     1     px(3)
              0   0     0     1];
end
```

```
% TRANSROTX Translation followed by rotation around
X axis
%
%        TRANSROTX(px,py,pz,alpha)    Returns    the
Homogeneous Transformation
%  of translation given in a vector p, followed by
rotation around X
%  axis,an angle alpha
%
%  T = transrotx(px,py,pz,alpha) assigns the result
(4x4 matrix ) to a
%  variable T
%  T = transrotx(p,alpha) same result but requires
p as row vector (size
%  1x3)
%
%    See  too  ROTXTRANS,  ROTYTRANS,  ROTZTRANS,
TRANSROTY, TRANSROTZ.
%
%  ***********************************************
%  * Developed by Wilmer Sanz F. Dec 2002        *
%  * The angle must be entered in degrees        *
%  ***********************************************



function tr = transrotx(px, py, pz, alpha)

if nargin == 4,
    % Conversion de grados a radianes
      a=alpha*pi/180;
    % Conversion of degrees to radians
           ca = cos(a);
           sa = sin(a);

      tr = [1     0     0     px
      0     ca    -sa   (py.*ca-pz.*sa)
      0     sa    ca    (py.*sa+pz.*ca)
      0     0     0     1];
elseif nargin == 2,
```

```
    % Conversion de grados a radianes
      a=py*pi/180;
    % Conversion of degrees to radians
            ca = cos(a);
            sa = sin(a);

      tr = [1     0     0     px(1)
      0     ca    -sa   (px(2).*ca-px(3).*sa)
      0     sa    ca    (px(2).*sa+px(3).*ca)
      0     0     0     1];
end
```

```
% TRANSROTY Translation followed by rotation around
Y axis
%
%     TRANSROTY(px,py,pz,phi)      Returns      the
Homogeneous Transformation
%     of translation given in a vector p, followed
by rotation around Y
%     axis, an angle phi
%
%     T  =  transroty(px,py,pz,phi)  assigns  the
result (4x4 matrix ) to a
%     variable T
%     T = transroty(p,phi) same result but requires
p as row vector (size
%    1x3)
%
%     See  too  ROTXTRANS,  ROTYTRANS,  ROTZTRANS,
TRANSROTX, TRANSROTZ
%
%  ***********************************************
%  * Developed by Wilmer Sanz F. Dec 2002         *
%  * The angle must be entered in degrees         *
%  ***********************************************
function tr = transroty(px, py, pz, phi)

if nargin == 4,
    % Conversion de grados a radianes
      f=phi*pi/180;
    % Conversion of degrees to radians
            cf = cos(f);
            sf = sin(f);

   tr = [cf 0     sf    (px.*cf+pz.*sf)
              0   1     0     py
            -sf   0     cf    (pz.*cf-px.*sf)
              0   0     0     1];
elseif nargin == 2,
    % Conversion de grados a radianes
      f=py*pi/180;
    % Conversion of degrees to radians
```

```
            cf = cos(f);
            sf = sin(f);

   tr = [cf 0      sf     (px(1).*cf+px(3).*sf)
             0  1      0       px(2)
           -sf  0      cf      (px(3).*cf-px(1).*sf)
             0  0      0       1];
end
```

```
% TRANSROTZ Translation followed by rotation around
Z axis
%
%     TRANSROTZ(px,py,pz,theta)     Returns     the
Homogeneous Transformation
%     of translation given in a vector p, followed
by rotation around Z
%     axis,an angle theta
%
%     T  =  transrotz(px,py,pz,theta)  assigns  the
result (4x4 matrix ) to a
%     variable T
%         T  =  transrotz(p,theta)  same  result  but
requires p as row vector
%     (size 1x3)
%
%     See   too   ROTXTRANS,   ROTYTRANS,   ROTZTRANS,
TRANSROTX, TRANSROTY.
%
%  ************************************************
%  * Developed by Wilmer Sanz F. Dec 2002         *
%  * The angle must be entered in degrees         *
%  ************************************************
function tr = transrotz(px, py, pz, theta)

if nargin == 4,
    % Conversion de grados a radianes
      t=theta*pi/180;
    % Conversion of degrees to radians
            ct = cos(t);
            st = sin(t);

   tr = [ct -st   0     (px.*ct-py.*st)
         st  ct   0     (px.*st+py.*ct)
              0   0     1     pz
              0   0     0     1];
elseif nargin == 2,
    % Conversion de grados a radianes
      t=py*pi/180;
    % Conversion of degrees to radians
```

```
            ct = cos(t);
            st = sin(t);

   tr = [ct -st   0     (px(1).*ct-px(2).*st)
         st  ct   0     (px(1).*st+px(2).*ct)
              0   0     1      px(3)
              0   0     0      1];
end
```

```
%    MPASH Matriz de Paso Homogénea para un
elemento de una cadena
%    cinemática
%
%    MPASH(theta,d,a,alpha) retorna la Matriz de
paso Homogénea.
%    theta: es el ángulo que gira el eje Xi para
estar paralelo a Xi-1,
%    alrededor de Zi-1
%    d: es la distancia que se desplaza el sistema
Si-1 para alinear el
%    eje Xi-1 con Xi
%    a: es la distancia que se desplaza el sistema
Si-1 para que su
%    origen coincida con el del sistema Si
%    alpha: es el ángulo que gira el eje Zi-1 para
alinearse con el eje
%    Zi, alrededor de Xi
%
%    *******************************************
%    * Creado por Wilmer Sanz F. Ene 2003       *
%    * Los ángulos deben introducirse en grados *
%    *******************************************

%    MPASH Transition homogeneous matrix for an
element of a kinematic
%    chain
%
%    MPASH(theta,d,a,alpha) returns the Transition
homogeneous matrix.
%    theta: angle around Zi-1 from Xi-1 to Xi1
%    d: offset along Zi-1 from Xi-1 to Xi.
%    a: displacement along Xi, from the origin of
Si-1 of the origin of
%    Si
%    alpha: angle around Xi, from Zi-1 to Zi
%
%  *********************************************
%  * Developed by Wilmer Sanz F. Dec 2003        *
%  * The angles must be entered in degrees       *
```

```
%   ***********************************************

function Ai = mpash(theta, d, a, alpha)
% Conversion of degrees to radians and calculate of
sines and cosines
      alfa=alpha*pi/180;
      th=theta*pi/180;
      ca = cos(alfa);
      sa = sin(alfa);
      ct = cos(th);
      st = sin(th);
            Ai = [ct    -ca.*st     sa.*st a.*ct
                  st ca.*ct -sa.*ct a.*st
                  0     sa    ca    d
                  0     0     0     1];
```

```
%     INVMPH Invierte una Matriz de Paso Homogénea
%
%     INVMPH(A) retorna la Inversa de la Matriz de
Paso Homogénea A
%
%  *************************************
%  * Creado por Wilmer Sanz F. Ene 2002   *
%  *************************************
%     INVMPH   Inverts   a   Transition   homogeneous
matrix
%
%     INVMPH(A) returns the inverse of Transition
homogeneous matrix A
%
%  *************************************
%  * Developed by Wilmer Sanz F. Dec 2002 *
%  *************************************
function InvA = invmph(A)

if A(4,1) ~= 0
   error('This  is  not  a  Transition  homogeneous
matrix')
end
if A(4,2) ~= 0
   error('This  is  not  a  Transition  homogeneous
matrix')
end
if A(4,3) ~= 0
   error('This  is  not  a  Transition  homogeneous
matrix')
end
if A(4,4) ~= 1
   error('This  is  not  a  Transition  homogeneous
matrix')
end
nx = A(1,1);
sx = A(1,2);
ax = A(1,3);
px = A(1,4);
ny = A(2,1);
```

```
sy = A(2,2);
ay = A(2,3);
py = A(2,4);
nz = A(3,1);
sz = A(3,2);
az = A(3,3);
pz = A(3,4);
n = [nx ny nz];
s = [sx sy sz];
a = [ax ay az];
p = [px; py; pz];
nuevopx = n*p;
nuevopy = s*p;
nuevopz = a*p;

InvA = [nx ny nz -nuevopx
   sx sy sz -nuevopy
   ax ay az -nuevopz
   0 0 0 1];
```

GLOSSARY

A

ANTHROPOMORPHIC ROBOT: Named for its emulation of the hip, shoulder and human arm. It is the RRR type, with the first axis perpendicular to the other two. It is, by far, the most used configuration of Industrial Robots.

APROACH (*a*): Unit vector in the Z axis of the system that has its origin in the TCP of a robot. It indicates the direction in which the terminal tool of a robot faces a work plane.

C

CARTESIAN ROBOT: PPP Industrial robot that performs the own movements of the Cartesian coordinate system. It is the second configuration most commonly used industrial robots.

CILYNDRICAL ROBOT: RPP Industrial robot that performs the own movements, rotation and displacements, of the cylindrical coordinate system.

D

DEGREE OF FREEDOM (DOF): Each of the movements that can perform a robot, independently, through the changes of values of its joints parameters.

DENAVIT HARTENBERG PARAMETERS (DH): The four parameters that characterize the elements or links of the kinematic chain and the relative movement therebetween. They are: θ, the angle of rotation in revolute joints (it is constant in prismatic joints), d, linear displacement in the prismatic joints (that is constant in revolute joints); a, magnitude of the line simultaneously normal or perpendicular to the axes comprising to each link; and α, torsion angle of each element. Their determination is the base of kinematic studies in industrial robotics.

DH: Acronyms of Denavit-Hartenberg and identification of parameters and algorithm that allow describing and modeling the kinematics of industrial robots.

DIFFERENTIAL KINEMATICS: Study of translational and rotational speed of the end effector of a robot, based on speeds of joints parameters.

E

EULER ANGLES: All possible combinations of consecutive simple rotations. In total they amount to 24 and the most known of them consist on the sequence of turns around the XYZ axes, known as RPY angles (Roll, Pitch, Yaw). They are a means to describe the changes in the orientation of the terminal tool.

F

FORWARD KINEMATICS: Study of changes in the position and orientation of the end effector of a robot, based on changes in the values of the joints parameters.

G

GENERALIZED COORDINATES SYSTEMS: Each of the coordinate systems drawn into the joint end in each of the elements or links in the kinematic chain. They are considered "firmly attached" to the link at the end of which are drawn. Describing his movement relative to the coordinate system that precedes, through a Transition Homogeneous Matrix, you are describing the movement of the link.

H

HOMOGENEOUS SPACE: Vector space where a vector can be represented by adding a dimension which is an integer factor called a scale factor. It also allows the representation of a matrix by adding a row and a column. It is useful in robotics, to apply Transformations (Homogeneous Transformations), which describe the changes in the position and orientation of a vector simultaneously by a single mathematical operation.

HOMOGENEOUS TRANSFORMATION: Representation in Homogenous Space of rotation matrices and translation vectors, which mathematically model the changes in the orientation and position of the robot terminal tool.

I

INDUSTRIAL ROBOT (IR): Universal designation of robotic manipulators used in industry. It is a definition made by various international organizations and federations, which keeps small differences between each of its versions. Here it is quoted the definition made by the International Federation of Robotics (IFR): "An automatically controlled, reprogrammable, multipurpose manipulator programmable in three or more axes, which may be either fixed in place or mobile for use in industrial automation applications".

INITIUM ROBOT: Name (alluding to "the beginning") given to 5R robot with ability of displacement, which was developed between the Post-grade of Instrumentation of the Faculty of Sciences of the Central University of

Venezuela and the University Institute of Technology Experimental "The Victoria". Its creator is Dr. Jesús Pérez.

INVERSE KINEMATICS: Study of changes required in the joints parameters for the end effector of a robot reaches a desired position and orientation.

J

JACOBIAN (Matrix): In general is a matrix where all elements are partial derivatives of functions of multiple variables. In Robotics the functions to lead, respect to time, they are describing the position and orientation of the end effector of a robot, so that the Jacobian matrix is the tool to describe the instantaneous velocities (translational and rotational) of the end effector according to speeds and instantaneous values of articular parameters (joints parameters). Its determination and evaluation is the heart and core of Differential Kinematics.

JACOBIAN (The): The name Jacobian also refers the determinant of the Jacobian matrix. In Robotics is used to determine singularities in the paths of the robots and the limits that define their Workspaces.

L

LOCATION (SITUATION): In Robotics; name given to refer together the position and orientation of the terminal tool of a robot.

N

NORMAL (*n*): Unit vector in the direction of X axis of the system that has its origin in the TCP of a robot.

O

ORIENTATION: Condition of the end effector of a robot that determines the way it approaches the working plane (*approach*) and sense of operation (*slide*). Its changes are modeled by rotation matrices or through the Euler angles.

P

PERSPECTIVE VECTOR: Row Vector located in the fourth row of the transformation matrix (T) of a robot. In Robotics, it is null. In computer vision applications (Machine Vision), it is not null and forms the basis of a homogeneous transformation called perspective transformation.

PITCH (θ): One of the RPY angles. It corresponds to the ascent rotation and descent of the human wrist.

POSITION VECTOR: Vector defined between the origin of the reference system of the robot base and the Tool Center Point. It is equivalent to position of TCP.

PRISMATIC JOINT (P): Type of joint corresponding to a linear movement or displacement.

PUMA ROBOT: *Programmable Universal Machine for Assembly*, or *Programmable Universal Manipulation Arm.* 6R Manipulator developed by Universal Automation (UNIMATION) in 1978. Its patent belongs to

Victor Scheinman. It has been used as model for developing of the most used anthropomorphic robots.

R

REVOLUTE JOINT (R): Type of joint corresponding to a turning movement.

ROLL (φ): One of the RPY angles. It corresponds to the human wrist rotation along a longitudinal axis extended through the forearm.

ROTATION SUBMATRIX: The first 9 elements (upper left corner) of the transformation matrix of a robot. Describe changes in orientation. His first column corresponds to the components of the vector n, the second column corresponds to the components of the vector s and the third, with the vector a.

S

SCALE FACTOR: Non-zero integer that is used as a basis for the representation of vectors and matrices in homogeneous space.

SCARA ROBOT: Toroidal Robot RRP, with parallel axes in the first two revolute joints. SCARA means Selective Compliant Assembly Robot Arm. It has become the third configuration of industrial robots most used.

SCOPE: Space formed by all points to where you can wound the end effector of a robot. It is another name for Workspace or Work Volume.

SLIDE (*s*): Unit vector in the direction of Y axis of the system that has its origin in the TCP of a robot. It represents the direction of work of the terminal tool.

SPHERICAL ROBOT: RRP Industrial Robot that performs the own movements of the Spherical coordinate system. Unlike to a SCARA Robot, its first two joints are perpendicular.

STANFORD ROBOT: 6R Robot designed by Victor Scheinman in the Laboratory of Artificial Intelligence of Stanford University, in 1969. It was the predecessor of PUMA Robot.

T

TOOL CENTER POINT (TCP): Central point of the tool or end effector of a robot. Its position on the initial reference system (S_0 or base) is usually generalized as "the position of the robot."

TRANSFORMATION MATRIX T: It is a homogeneous matrix to describe the changes in the position and orientation of the terminal tool of a kinematic chain. It includes a 3x3 submatrix describing the orientation (comprising the vectors n, s and a), a null row vector (perspective), a column vector describing the position of the TCP and the unit element corresponding to the Scale Factor. It is obtained as the result of the post-multiplication of Transition Homogeneous Matrices.

TRANSITION HOMOGENEOUS MATRIX: Each of the matrices that model the relative movement between consecutive links in the kinematic chain. They are obtained from the DH parameters.

U

UNIMATE ROBOT: The first Industrial Robot (1961) developed by UNIMATION Inc. (founded by George Devol and Joseph Engelberger) and used to work with heated die-casting machines in General Motors. Its weight was near to 1800 Kg (4000 lb).

W

WORK VOLUME or WORKSPACE: See Scope.

Y

YAW (ψ): One of the three RPY angles. It corresponds to the restricted lateral rotation made by the human wrist about an axis which crosses perpendicularly from the top or zenith.

BIBLIOGRAPHY

Barrientos, A., Peñín L., Balaguer C. & Aracil R. (1997). ***Fundamentos de Robótica***. Mc Graw Hill. Madrid

Baker M. (2008). ***Math– Euler Angles***. Retrieved from:

http://www.euclideanspace.com/maths/geometry/rotations/euler/index.htm.

Carnegie Mellon ***Robot. Hall of Fame. The 2003 Inductees: UNIMATE***. Retrieved from: http://www.robothalloffame.org/unimate.html.

Coiffet, P. y Chirouze, M. (1983). ***An Introduction to Robot Technology***. Mc Graw Hill. Paris.

Calero, Roque y Carta José (1999). ***Fundamentos de mecanismos y máquinas para ingenieros***. Mc Graw Hill. Madrid.

Denavit, J. & Hartenberg, R. (1955). A kinematic notation for lower-pair mechanisms based on matrices, Transactions ASME Journal of Applied Mechanics: 23, 215–221.

Donald, S. y Dunlop, G. (2001). ***Retrofitting Path Control to UIMATE 2000B Robot***. Retrieved from: http://www.araa.asn.au/acra/acra2001/Papers/Do

nald.pdf. Proceedings 2001, of the Australian Conference of Robotics and Automation.

Fu, K., González, R. & Lee, C. (1994). ***Robótica. Control, Detección, Visión e Inteligencia***. Mc Graw Hill. México.

Giamarchi, Frédéric (2001). ***Robots Móviles. Estudio y construcción***. Paraninfo. Madrid.

Infolab Stanford University. ***Robots and their Arms***. Retrieved from:

http://infolab.stanford.edu/pub/voy/museum/pictures/display/1-Robot.htm).

Lagonell, H. & Lartíguez, N. (2004). ***Diseño y construcción de un Robot Cilíndrico de cuatro grados de libertad***. Available in Rental Library of the School of Mechanical Engineering of the University of Carabobo. Valencia, Venezuela.

Muñoz, M. (2008). ***Manual de Vuelo***. Retrieved from: http://www.manualvuelo.com/PBV/PBV15.html.

Paul, Richard (1981). ***Robots Manipulators***. The MIT Press. Massachusetts.

Rentería, Arantxa & Rivas, María (2000). ***Robótica Industrial. Fundamentos y aplicaciones***. Mc Graw Hill. Madrid

Rivas, Adolfo (2005). ***Diseño e Implementación de un Controlador en Interfaz Gráfica de usuario interactiva con computador***

personal, para un Robot Manipulador SCARA de dos grados de libertad, dedicado al marcaje de piezas en láminas metálicas. Available in Rental Library of the School of Electrical Engineering of the University of Carabobo Valencia, Venezuela.

Sandin, Paul (2003). ***Robot Mechanisms and Mechanical Devices Illustrated***. Prentice Hall. Britain, UK.

Santos, José & Duro, Richard (2005). ***Evolución artificial y robótica autónoma***. Alfaomega. México D.F.

Sciavicco, Lorenzo & Siciliano, Bruno (1996). ***Modeling and Control of Robot Manipulators***. Mc Graw Hill. New York.

Sanz, Wilmer (2002). ***PGIBot Media***. [CD Rom]. University Central of Venezuela. Caracas.

Torres, F.; Pomares, J.; Gil, P. Puente, S. & Aracil, R. (2002). ***Robots y Sistemas Sensoriales***. Prentice Hall. Madrid.

ANALYTICAL INDEX

Made in United States
Orlando, FL
04 November 2026

90200405R00140